Multi-ethnic Bird Guide

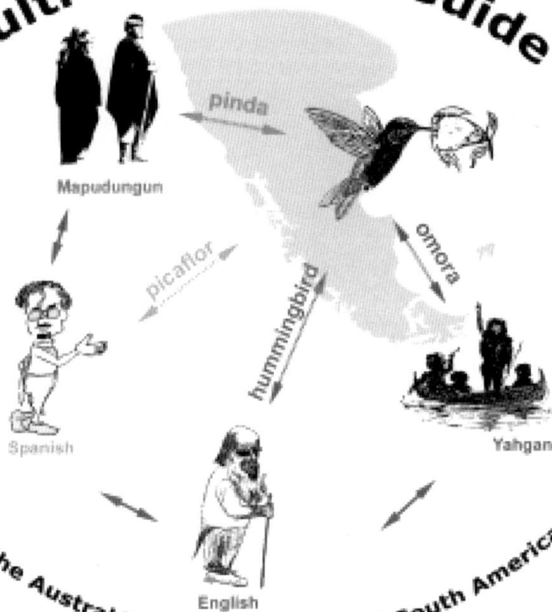

of the Austral Temperate Forests of South America

Multi-ethnic Bird Guide

of the Austral Temperate Forests of South America

Editors
Fantástico Sur

Multi-ethnic Bird Guide
of the Austral Temperate Forests of South America

General Direction and texts
Ricardo Rozzi

Ethnographic research and stories collection
Francisca Massardo & Ricardo Rozzi

Ornithological research
Ricardo Rozzi, Christopher Anderson, Steven McGehee,
George Clark, Guillermo Egli & Eduardo Ramilo

Yahgan bird names and stories
Úrsula Calderón & Cristina Calderón

Mapuche bird names and stories
Lorenzo Aillapan

Biographies
Cristina Zárraga & Lorenzo Aillapan

Spanish text review for recordings
Mapuche stories: Francisca Massardo & Christopher Anderson
Yahgan stories: Cristina Zárraga & Francisca Massardo

Narrators of bird common names recorded in the CDs
Spanish: Francisca Massardo, Luis Gómez & Lorenzo Aillapan;
English: Steven McGehee & Lorena Peñaranda

Narrators (CDs)
Yahgan: Úrsula Calderón & Cristina Calderón
Mapudungun: Lorenzo Aillapan
English: Kurt Heidinger, Uta Giesen & Steven McGehee

Field recordings
Birds: Guillermo Egli, Steven McGehee, Ricardo Rozzi & Nolberto González
Bird names & stories: Ricardo Rozzi, Steven McGehee & Nolberto González

Recording design and sound production
John Schwenk & Nolberto González

Photography
Oliver Vogel & others

Logistic support
Ezio M. Firmani & Francisca Massardo

English Version
Translation from Spanish: Christopher Anderson
Revised edition: Kurt Heidinger & Ricardo Rozzi

Editors:

Fantástico Sur - Birding & Nature
Magallanes 960 2nd Floor
P.O. Box 455, Punta Arenas - Chile
Phone/Fax +56 61 247194
info@fantasticosur.com
www.fantasticosur.com

Registration N°134.984, year 2003
ISBN 956-8007-06-7

Graphic Design, Layout
y Photographic digitalization:

Fantástico Sur / Miguel Angel Montecinos

Cover and backcover

Design: Miguel Montecinos y Ricardo Rozzi
Photographs: Oliver Vogel (Lenga forest, Cristina Calderón and Lorenzo Aillapan), John Schwenk (Úrsula Calderón and Ricardo Rozzi), Alejandro Kusch (*Campephilus magellanicus*), Steve Morello (*Glaucidium nanum y Scytalopus magellanicus*), Eduardo Pavez (*Buteo polyosoma*), and Ricardo Stanoss (*Ceryle torquata*).

PRINTED IN SANTIAGO, CHILE

INDEX

PREFACE

The forests of southern Chile and Argentina are unique ecosystems and they are highlighted by their great richness and productivity. A relevant element of their diversity is the avian fauna, which are very close animals for humans; they are in our dreams and our fears, in myths and legends. They are in the stories that our grandparents pass down to us, and they are in the national coat of arms of many nations (e.g. the Andean condor is on the shield of Chile, Bolivia, and Colombia). With their calls and cries, with their flagrant presence, they are part of our daily life. Of course, the birds are in the forest, walking on the soil, jumping between the bushes, clambering along the trunks and flying over the canopy. Birds have an important role in the maintenance and function of the forest. There are species that excavate the soil. There are those that disperse seeds, others that consume insects and those that feed upon dead animals. Birds tell us about the health of the forest, and there are some "accusers" that disappear when the trees are cut, with their absence indicating to us that this environment is losing some of its qualities.

The reforms that are being developed now in Chile's education system have as their objective the incorporation, as a fundamental element of learning, the values of the environment that surrounds the schools, consequently giving the student the opportunity for a concrete experience. In such context, this book is an invitation to look, to see and to listen. It is a call to the young people, from ages 1 to 99, to be entertained and to learn. It is a call to know about bird life and by way of them to understand the forest; because understanding the forests, and with them nature, we all will be able to have a better life.

Juan C. Torres-Mura
Ex-President, Union of Chilean Ornithologists
Curator of Ornithology, National Natural History Museum
Santiago, Chile

August 2001

INTRODUCTION

Ricardo Rozzi

I. Collective Work

The bird songs, names and stories recorded on the CDs of this guide book begin with the song of the hummingbird, the cry of the barn owl, the percussion and calls of a woodpecker family, together with the murmur of the Robalo River and the music of Mapuche poet Lorenzo Aillapan's *trutruk a*. The auditory journey continues with the vocalizations of fifty birds that inhabit the forests of austral South America along with their Yahgan, Mapudungun, Spanish and English names. Then, we present seventeen Mapuche stories, ten Yahgan stories and one Mapuche-Yahgan story of the austral thrush, considered the orchestra director of birds and austral nature. The journey ends with the song of the hummingbird, the austral thrush, and other birds, together with the wind and music of *Gli Uccellini*, a musical piece for the piccolo by Chilean composer Andrés Alcalde. As a whole, the recordings of this guide book express the voices of multiple species and indigenous, rural and urban cultures, whose lives are interwoven in the temperate forest region of South America.

The recordings also manifest a collective character in the making of this guide, given that the varied voices involve many different birds and people. The collective character of the bird songs and narratives registered in this book is even greater if we consider that these come from evolutionary processes and oral traditions. The birds have sung in the austral forests for millions of years; the Mapuche and Yahgan peoples have transmitted their bird stories from generation to generation for hundreds or thousands of years. These voices, then, inherit a long co-evolutionary history between the diverse birds and human beings that have co-inhabited, and still co-inhabit, the forested regions of austral South America. The records of this bird guide offer us today a biocultural gift for those who yearn to hear, understand and respect the voices of the distinct birds and human beings with whom we share our lives in this remote part of the planet.

The necessity of creating this multi-ethnic bird guide about the bio-cultural diversity associated with the avifauna of the South American temperate forests arose during a workshop conducted in 1995 in the Cucao Sector of Chiloé National Park, in the territory of the Huilliche communities of Chanquín and Huentemó. The shared activities and conversations with the park rangers and ornithologists of southern Chile and Argentina expressed not only how rich the bird diversity is, but rather also the appreciations that diverse ornithologists have regarding them. I thought then that the biological diversity of the birds and

Ornithologists and park rangers from southern Chile and Argentina observe birds captured in mist nets during a workshop entitled "Recognizing the Forest Birds of Chiloé." The making of this multi-ethnic bird guide was motivated by this workshop, organized by the Senda Darwin Foundation, CONAF – Chiloé, and the Chilean Union of Ornithologists in the Cucao Sector of Chiloé National Park in December, 1995.

the cultural diversity of our observations, names and stories about them could be known and understood better by way of preparing a multi-ethnic bird guide.

The departure point was the preparation of a sound register for the bird's vocalizations, a photograph archive, natural history texts and species distribution maps, considering the birds' seasonal movements. This would permit: i) the visual and auditory identification of each species, ii) the acquaintance with their distributions, and iii) a synthesis in simple language of our natural history knowledge from field work and bibliographic accounts regarding forest birds.

At the same time, we began to conduct interviews in the field, and an extensive ethnographic bibliography review. The interviews initiated in the Chanquín and Huentemó communities in 1995 included later Manao, Linao and Compu on Chiloé Island, Galletué and Icalma in Alto Bío-Bío and Nahueltoro in Ñuble. The conversations during this slightly erratic itinerary took us to the Mapuche poet Lorenzo Aillapan, an *Ünümche* or Mapuche Birdman. It was working with the Master Aillapan that permitted us later to give a more comprehensive understanding to the interviews and intensive Mapuche ethno-ornithological bibliography review.

Mapuche poet Lorenzo Aillapan (center), recording Mapuche bird stories in the forests of the Omora Ethnobotanical Park, Isla Navarino, with Steven McGehee (right) and Ricardo Rozzi (left).

The course of searching for understanding how birds and forest ecosystems of southern South America are perceived by distinct cultures lead me to develop a doctoral dissertation in ecology, combined with a Master´s in philosophy, which would include as well the study of Yahgan traditional ecological knowledge. As part of this larger project, we invited a team of scientists, sound engineers and teachers, both Chilean and international, with whom we traveled to Isla Navarino, Cape Horn County, in January 2000.

The Yahgan grandmother Úrsula Calderón during a recording session with Ricardo Rozzi in Bahía Mejillones, Isla Navarino.

Living together with the Indigenous Yahgan Community of Bahía Mejillones and encountering with the wisdom of the Yahgan grandmothers Úrsula Calderón and Cristina Calderón, permitted us to come closer to traditional Yahgan ornithological knowledge. During the first year of work on Isla Navarino, the memories of the grandmothers Úrsula and Cristina, concerning the Yahgan bird names and narratives, initially were flowering rapidly and then later more gradually. I will never forget the moment when we had given the project up as finished and put away the recording equipment, when Grandmother Úrsula knocked on the door saying, "I remembered: *wichoa*, that is the Yahgan name of the red-breasted one." That was the last name registered in this guide book.

In the CD recordings that accompany this book only the names of the birds that were remembered by the grandmothers Úrsula and Cristina are recorded. Other Yahgan names appear written at the beginning of the texts for each species. With respect to the narratives, we adopted a different approach. After recording all of the stories that were remembered by the grandmothers, we decided together with teacher Luis Gómez (grandson of Grandmother Cristina) "to retell" to Úrsula and Cristina four Yahgan birds stories that possess a great educational value. These stories about the Magellanic Woodpecker, Andean Tapaculo, Buff-Necked Ibis and Greenback Firecrown Hummingbird had been told by their father Juan Calderón and other members of the Yahgan community to anthropologist Martin Gusinde between 1920 and 1923. Such "intervention" had as its objective the stimulation of their memories and to compose this guide book as educational material that would contribute to the continuity of the

The recordings of the bird names and narratives in Yahgan, Mapudungun, Spanish and English were done in the forests of the Yahgan territory of Mejillones Bay and in the Omora Ethnobotanical Park, Navarino Island, between January 2000 and May 2001. In the photograph (from left to right), ethnobiologist Francisca Massardo, Yahgan Grandmother Úrsula Calderón, ornithologist Steven McGehee, Mapuche poet Lorenzo Aillapan, Yahgan Grandmother Cristina Calderón, and teacher Luis Gómez (grandson of Grandmother Cristina) in Mejillones Bay, February 2001.

Yahgan language and traditional ecological knowledge. In the case of the stories of the ibis, or *lejuwa*, and the woodpecker, or *lana*, the grandmothers recuperated a more or less living memory. In the case of the hummingbird, or *omora*, they only recalled the name of *omora* as was told by Grandmother Julia. For the story of the tapaculo, or *tuto*, they did not evoke any remembrance, and this story was recorded by Luis Gómez with his grandmother Cristina based on the text registered by Martin Gusinde.

The first phase of work making the audio recordings culminated in February 2001, when the Mapuche poet Lorenzo Aillapan accepted our invitation to come to Isla Navarino. During the summer, with the Grandmothers Ursula and Cristina, Lorenzo, and the whole team, we recorded the names and stories of the birds in Yahgan, Mapudungun, Spanish and English, seated around the fire or under the canopy of the austral forests.

Ecologist Christopher Anderson prepared the English version of the Mapuche stories, working directly with the poet Lorenzo Aillapan and of the Yahgan stories with the

Writer Kurt Heidinger (left) and sound engineer John Schwenk (right) record and edit the sound files in the studios of WHUS Radio (University of Connecticut, USA) in order to generate the counterpoint between the diverse languages of birds and humans in the English version of the CDs in this guide book.

ethnobiologist Francisca Massardo. Christopher's texts were then edited by writer Kurt Heidinger, who also recorded many of the stories in English. Other stories and names in English were recorded by Steven McGehee, Uta Giesen and Lorena Peñaranda, which allowed the incorporation of variations in tone and pronunciation in this language, which acquired importance at the austral extreme of the Americas with the arrival of the Anglican missionaries 150 years ago.

The Spanish version of the Mapuche stories was recorded by the poet Aillapan with their onomatopoeic imitations. The Yahgan stories were recorded in Spanish first by the grandmothers Úrsula and Cristina Calderón. However, in April 2001 poet and writer Cristina Zárraga, who is the granddaughter of Grandmother Cristina, arrived to Isla Navarino. Cristina Zárraga agreed to collaborate and revise, together with her grandmother, the Spanish texts of the Yahgan narratives. Then, the

Writer Cristina Zárraga (left) and researcher Uta Giesen (center-left) record Yahgan stories with Ricardo Rozzi (center-right) in Spanish and English, while photographer Oliver Vogel (right) chronicles the scene on the north coast of Isla Navarino.

majority of the Yahgan stories were recorded in Spanish in her voice.

The meeting with Cristina Zárraga gave origin to a fertile relationship that is expressed in her own work as a writer and educator, as well as an investigator of Yahgan culture, which is continued to this day. With the poet Aillapan, the meeting also has permitted the continuation of joint work about the avifauna. Other fruit of this shared work was publishing the CDs and book *Twenty Winged Poems of the Native Forests of Chile*, which combined the poems of Lorenzo with the recordings of songs and illustrations prepared by a team of ornithologists. Today these poems continue their linguistic drift in the composition of twenty musical pieces by Andrés Alcalde and other Chilean composers. As such, following the oral tradition of the Yahgan and Mapudungun languages, this guide book has been a dynamic process that has given rise to new outlooks and songs in the dialogue with birds at the tip of South America.

Nolberto González, music teacher in Puerto Williams and director of Button Records, and Ricardo Rozzi edit the sound files of birds and human voices. The bird recordings were done by Nolberto González, Steven McGehee and Ricardo Rozzi on Navarino Island between January 2000 and March 2001. For the bird species that are found only in the more northern forests, Guillermo Egli's sound archives were used.

II. The Biocultural Mosaic of the Austral South American Temperate Forests

In order to facilitate the understanding of the relationship between birds, cultures and types of forested habitats, here we present a brief synthesis regarding the mosaic of forest ecosystems and cultures in southern South America. The ethnographic and forest-type maps will serve as a reference when viewing the distribution maps that appear together with each bird species. Combining these maps it is possible to deduce in what types of forests and in which Mapuche, Yahgan or other indigenous territories each bird inhabits.

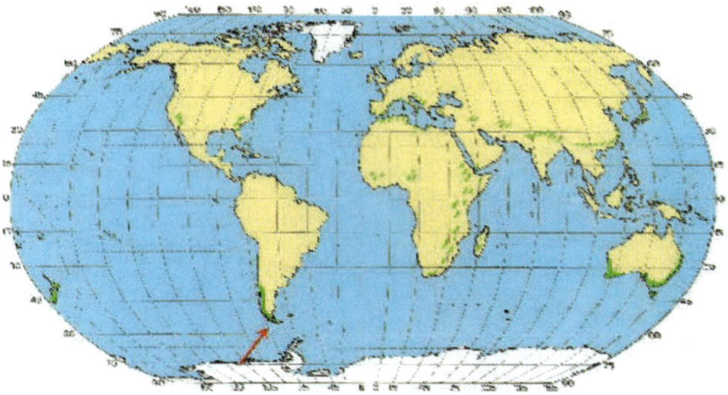

World map that illustrates the distribution of the world's temperate forests (in green). In contrast to tropical forests, the temperate forests are found at high latitudes; and, those of South America are the world's southernmost forested ecosystems (red arrow). (Figure modified from Fuentes 1994).

II.1 Biological diversity

The temperate forests of South America are the world's southernmost forested ecosystems. They extend for approximately 2000 km from central Chile (~35°S.) until the tip of the continent (56°S), and for a short distance along the eastern slope of the Andes Mountain Range in Argentina. In this guide book we have omitted the temperate forests of the highlands in southern Brazil. In order to make it clear that we are focused on the long and narrow strip of temperate forests from southern Chile and Argentina, we entitled this book the "Multi-ethnic Bird Guide of the Austral Temperate Forests of South America."

Forests dominated by Monkey Puzzle trees or pehuen (Araucaria araucana) *in the area of Alto Bío-Bío (38° S).*

The fauna and flora of the southern South American temperate forests possess an interesting mix of biogeographic origins, which include two divergent biotas: the Neotropical (warm) and the Gondwanic (sub-Antarctic, cool temperate). In addition, along the length of the extensive latitudinal strip of forests marked temperature and rainfall variations occur. A complex topography, which includes the Coastal Mountain Range, the Central

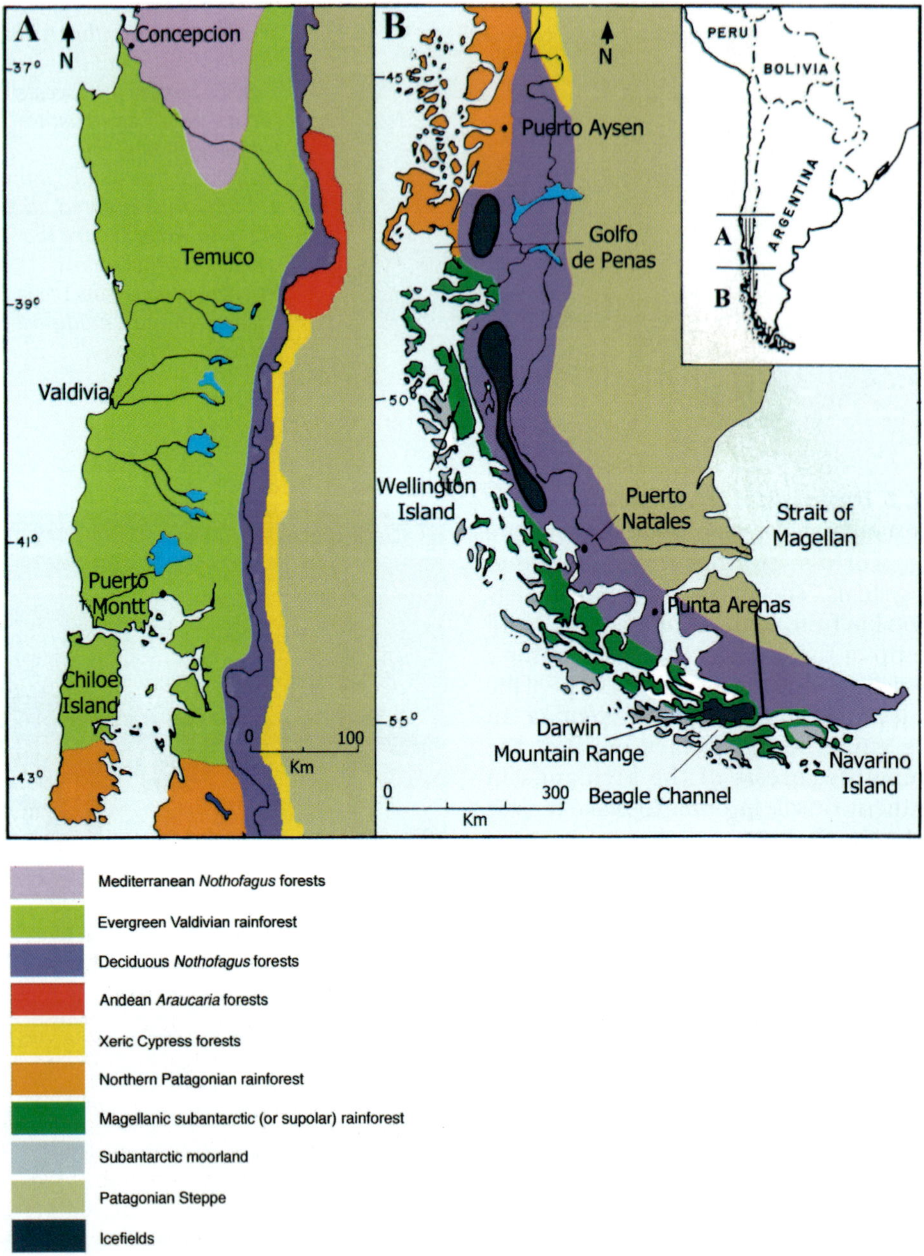

Main forest types in the region of the temperate forests in southern South America. The number and name of forest types vary with different classification systems, but there is no doubt about their marked diversity in the region .

Evergreen rainforests of the Valdivian ecoregion, with a complex canopy structure (left) and understory (above). Photographs taken in the Alerce Andino National Park (41.5ºS).

Valley, the Andes, and the southern Archipelago zone generates further environmental heterogeneity. The combined effect of climatic, topographic and biogeographic variations results in a remarkably diverse mosaic of forest ecosystems along the length of the latitudinal gradient.

The forest type mosaic includes landscapes as diverse as the high Andean zones, dominated by the *pehuen* or the Monkey Puzzle Tree (*Araucaria araucana*), the dense and structurally complex Valdivian rainforests, the Andean-Patagonian deciduous forests, and the sub-Antarctic (or subpolar) Magellanic rainforests.

Another determining characteristic of the southern forests' biological diversity is their biogeographical isolation, compared to other forest biomes.

*Deciduous forests dominated by Tall Deciduous Beech (*Nothofagus pumilio*) at the southern of its distributional range. Photograph taken in the Omora Ethnobotanical Park, Navarino Island (55ºS).*

*Sub-Antarctic forests on Isla Hoste, Cape Horn County. Low-lying areas are dominated by evergreen forests of Evergreen Beech (*Nothofagus betuloides*). In contrast, in higher zones deciduous forests dominated by Low Beech (*Nothofagus antarctica*) prevail. The latter are characterized by their bright red colors in autumn.*

*Close-up of the evergreen forests dominated by the Evergreen Beech (*N. betuloides*), which grow all the way to the high tide line along the coasts of fjords, channels and island landscapes that typify the Yahgan territory.*

Great orographic and climatic barriers separate the southern temperate forest biome from the nearest tropical forests, by at least 1500 km. To the north lies the Atacama Desert, one of the driest in the world. Westward and southward we find the Pacific Ocean. To the east, rise the high Andean Mountains and beyond them lies the vast, dry Patagonian steppe and other xeric plant communities in Argentina. In this way, the austral temperate forests of South America really are a biogeographic island. This insular character has promoted the evolution of a largely endemic biota. Close to 90% of the woody plant species are endemic. Among vertebrates, levels of endemism reach 80% for amphibians, 36% for reptiles, 33% for mammals and 30% for terrestrial birds. The latter percentage rises to 66%, if we consider the southern extreme of South America with the whole suite of terrestrial habitats. Hence, two-thirds of the bird species that you will be able to observe in the region only live in southern Chile and

Argentina, and half of these (that is, a third of all birds) are endemic to the forest habitats of the austral region. Such a high degree of endemism in the avifauna is much greater than levels recorded in other forested continental regions of the world; it is rather comparable to the high degrees of endemism found on oceanic islands.

II.2 Cultural diversity

The South American temperate forest biome has supported not only unique ecosystems and biota, but also a mosaic of idiosyncratic and diversified indigenous cultures. Among these Amerindian cultures were the world´s most southerly pre-Columbian inhabitants, the Yahgans. They occupy the archipelago territory of the extreme south of the Americas (55-56°S), located from the southern coast of Tierra del Fuego to Cape Horn. The Yahgans possess a long canoeing tradition and they host a refined traditional ecological understanding of the forests in the austral archipelago and their avifauna.

North to the Yahgan territory, Tierra del Fuego was inhabited by the Selknam or Ona and by the Haush. These were terrestrial hunter groups. North of the Strait of Magellan, the Patagonian steppe, which extends to the eastern border of the Andean-Patagonian forests, corresponds to the territory of another tribe of land hunters, the Aonikenk or Tehuelches. Along the western strip of these latitudes, the rainy region of the sub-Antarctic forests from Isla Wellington (49°S) to southwest Tierra del Fuego (54°S) are the territory of the Kaweskar or Alakaluf canoe culture. These southern Amerindian peoples host unique practices and ecological knowledge, associated with their capacity to inhabit areas with very rigorous climatic conditions. Their cultural singularity is due also to the degree of isolation that they experienced with respect to contact with other dominant South American cultures, such as the pre-Incan and Incan civilizations.

The northern half of the South American temperate forest region hosts the territories of the Mapuche people (35-42°S). The name Mapuche means "People of the Earth" (*Mapu* = Earth; *che* = people). In turn, Mapuche people speak Mapudungun, which means the "Language (=*dungu*) of the Earth." This culture possesses a vital link with its environment and a capacity to communicate with other beings, including birds, as will be possible to appreciate in the texts and the recorded names, stories and their onomatopoeias in the CDs in cluded in this book. Today, three groups of Mapuche people inhabit distinct forest types: the *Pehuenche* (people of the *Pehuen*), closely associated with forests dominated by Monkey Puzzle trees or *Araucaria araucana* (=*Pehuen*); the *Lafkenche* (people of the sea or *Lafken*) associated with the coastal forests and other marine ecosystems of central southern Chile; the *Huilliche* (people of the south or *Huilli*), associated with diverse types of evergreen rainforests in the Valdivian forest ecoregion between the Toltén River (38°S) and the south of Chiloé Island (42°S).

Today, cultural diversity requires even greater consideration and support than biological diversity. Half of the indigenous languages that were spoken in Chile are extinct, and a third of the extant ones are seriously threatened. At the world-wide level, it is also considered that the threats to linguistic and cultural diversity are even greater than those suffered by biological diversity. Indeed, 90% of the languages spoken today in the world could cease to exist during the 21st century! In this context, this multi-ethnic book aspires to contribute to the continuity of the Yahgan and Mapudungun languages, as much as to the conservation of the bird songs and their ecological interactions with the temperate forests in southern South America.

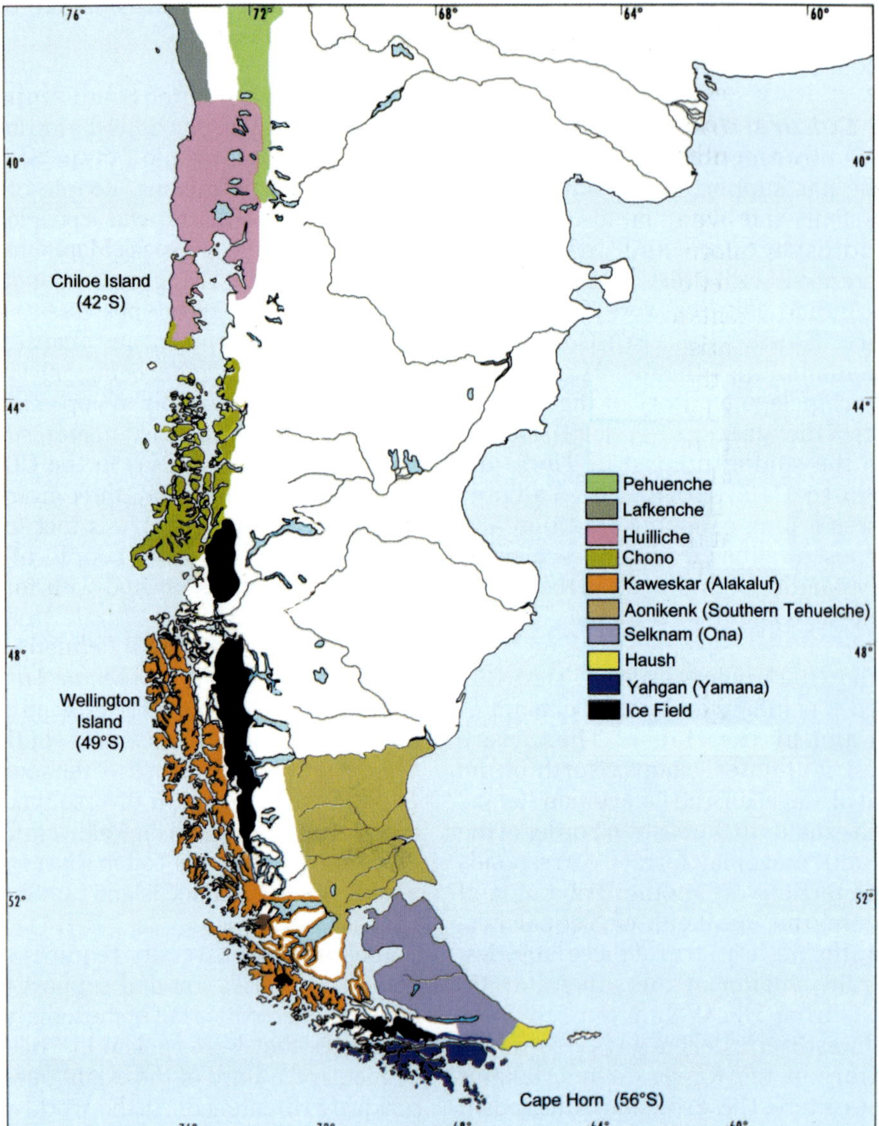

Ethnographic map of the temperate forest region of southern South America.

III. Ethno-ornithological Philosophy and Environmental Ethics

This guide is an invitation to wander through the forests of southern South America and listen to the voices of the birds and people that inhabit them. It provides an orientation in order to perceive the uniqueness of the ornithological voices or expressions that emanate from each ravine, bay or slope in the world's southernmost forests.

Each time that we observe a bird, our perception is informed by something of the bird, and something of ourselves—our senses, instruments and concepts. In this way, a dialogue arises between the birds themselves and the names and stories that express our human perceptions of them. The variety of birds is as great as the richness of the ornithological narratives with their particular ways of looking at the birds and the world. This book guides us to recognize the cultural and biological diversities, and to understand the relationships between both diversities in the Mapuche, Yahgan and scientific stories.

Yahgan and Mapudungun are languages with an oral tradition, and within the forests it is more common to hear than to see birds. Therefore, sensitive listening helps to cultivate our intelligence in the recognition of birds, their names and stories. For this reason, we suggest beginning by listening to the recordings of the CDs that accompany this guidebook. Familiarize yourself with the birdcalls and their names in Yahgan, Mapudungun, Spanish and English. Each name, each sound, has a story; you will hear, for example, that the species *Sephanoides sephaniodes* is called:

Upon familiarizing yourself with the audio and recognizing the calls and names of the birds, you will be able to discover how the names and stories show us complementary facets about the birds, cultures, forested ecosystems and their interactions in the austral extreme of the American Continent. For example, for the species Sephanoides sephaniodes the names in Mapudungun and English allude to the rapid, beating sound of the wings. Pinda or pigda refers to the sound that is produced when one things is rubbed against another (=pigudcun). In English this particular sound produced by the beating of the wings of this bird is called humming, thus the name hummingbird. The names in Spanish and Yahgan allude, in contrast, to the behavioral aspects that can be observed in this bird. Picaflor denotes the habit of visiting or "piercing" (picar) the flowers to drink their nectar. The valiant character possessed by omora, the tiny warrior in the Yahgan narratives, could be related to its territorial behavior of defending the flowers that it visits in order to feed itself.

In the face of the diversity of names and stories about each bird species, a question arises: How can there exist so many narratives about the same bird? Are they all false or all true? Are some truer than others? In order to respond to these questions, this multi-ethnic guide book helps to interpret the traditional stories from the viewpoint of ecology, including the natural history and life history of the birds, as well as evolutionary and human ecology.

As a point of departure in the ecological outlook for listening and reading the ethno-ornithological stories, we propose the following analogy:

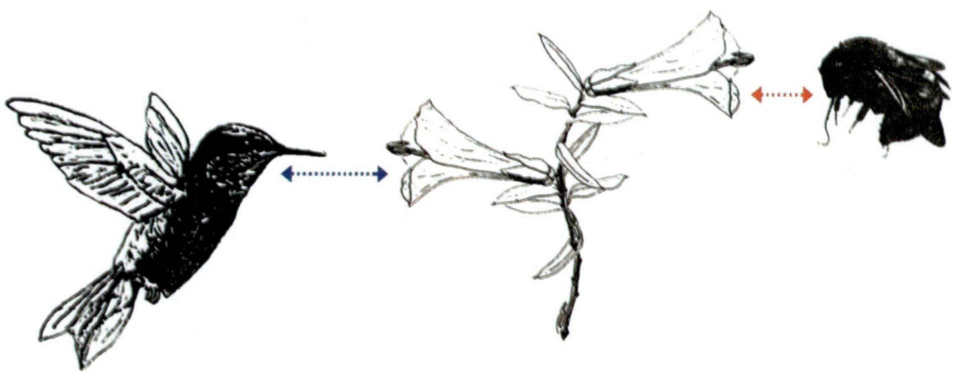

*In the temperate forests of southern South America, the hummingbird (*Sephanoides sephaniodes*) and the bumblebee (*Bombus dahlbomii*) visit the flowers of the kolkopiw vine (*Philesia magellanica*). The hummingbird and the bumblebee take the nectar of this flower, obtaining the sugar and water that permit them to survive. On the other hand, upon visiting the flowers, the hummingbird and bumblebee help the reproduction of this plant by transporting pollen from flower to flower. In this way, more flowers*

will be produced and the cycle of life continues. This scene can be observed in the native forests of southern Chile and Argentina—and it appears "natural." It would not occur to us that the bumblebee *should teach the* hummingbird *to visit the* kolkopiw *flowers in the way of an insect—or vice-versa: the hummingbird should teach the bumble bee to visit the flowers in the way of a bird. Each species visits the flowers in its own manner and both survive in the austral ecosystems.*

In an analogous way, we can interpret the diversity of Yahgan, Mapuche and Western-scientific viewpoints about birds. Each of these cultures, and languages, recognizes the bird species and interacts with them in a particular way. As has occurred with the hummingbird and bumblebee, the Mapuche and Yahgan cultures have inhabited and survived in the austral forests for thousands of years. As a result, the traditional ecological knowledge (including the recognition of the birds and their relationship with the biological communities and ecosystems) has been appropriate to permit the survival of these people. As a result, their stories are not merely "myths" or "legends," but rather they interpret the natural reality in an effective manner.

The traditional ecological knowledge contained in the Yahgan and Mapuche ornithological narratives permits us to expand our ways of knowing about, and inhabiting, nature, and of living together with the birds and their ecosystems. This enriches our experience of visiting and exploring the austral forests of South America. At the same time, the broad array of forms of traditional ecological knowledge offer us valuable perspectives regarding the conservation of these ecosystems and approaches to inhabiting them in a sustainable manner.

The Mapuche and Yahgan ornithological stories not only contrast scientific views; we also find substantial similarities between them. For example, the indigenous narratives share two central notions with the contemporary, ecological-evolutionary perspective: 1) the sense of kinship between human beings and birds, derived from common genealogies or evolutionary histories, and 2) the sense of biotic communities or ecological networks, of which humans and birds form part.

III.1 Our relatives, the birds

With respect to the evolutionary sense of kinship, modern biology has discovered that the human species (*Homo sapiens*) possesses cells, blood vessels, vertebrae and other structures that are very similar in birds. As can be observed on the cover of this book, the eyes of birds and human beings are very similar, as well. When walking through the forests, use your binoculars to discover how similar to your own eyes are to the eyes of the birds. This type of similarity would be explained evolutionarily by the existence of a common origin or ancestor for birds and mammals. Such a common ancestor would have existed millions of years ago. In the broad sense of a common evolutionary history and biological nature, these affirmations that result from a scientific analysis share essential concepts with those contained in the Mapuche and Yahgan narratives, such as the sentence "in ancestral times when birds were humans" with which several Yahgan stories begin. Let's consider the following example:

One of the Mapuche stories tells that Lautaro or Leftraru, the greatest of Mapuche warriors, descended from the lineage of eagles and caracaras or traru (Caracara plancus)*. So much so, that the name Leftraru means fast (=lef) traru.*

One of the Yahgan stories affirms that the woodpeckers, or lana, (Campephilus magellanicus) *are descendants of a pair of Yahgan siblings.*

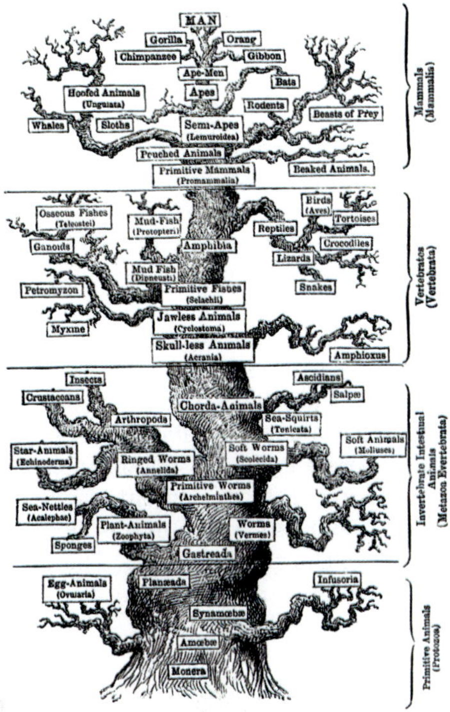

The tree of life by German biologist Ernst Haeckel illustrates the scientific-evolutionary theory of Charles Darwin, which proposes that human beings possess a common evolutionary origin shared with all living beings.

The genealogies of birds and humans illustrated in the former example allude to an evolutionary history that implies a common nature for bird species and the human species. From the point of view of contemporary environmental ethics, the three cultural perspectives—Mapuche, Yahgan and scientific— emphasize the *intrinsic value* of avifauna because *the birds are our distant evolutionary relatives*. This implies that, to some degree, the existence of birds can be subject to moral considerations based on ontological and ethical judgments on par with those we use to judge the value of human life.

III.2 Ecological and ethical networks

Regarding the concept of ecological networks, the participation of the human species in food webs and other biotic interactions is evident today for contemporary science. Human beings are not separate from nature; we constitute a keystone species for many ecosystems. We influence ecosystem processes, at the same time that ecosystems provide us with goods (such as medicinal plants, fishes and other food) and services (such as clean air, the regulation of climate and hydrologic cycles, increasing soil fertility, and the control of pests). The new discipline of ecological economics has determined, for example, that the cheapest and most sustainable method to assure the supply of drinking water is the protection of forested watersheds. Concepts such as these, which are recent for science, are rooted in the Mapuche and Yahgan traditional ecological knowledge.

Yahgan stories —such as that of the Andean Tapaculo or *tuto* (*Scytalopus magellanicus*), which like the ancient Yahgans constructed canoes of bark and fibers— and Mapuche narratives —such as that of the Meadowlark or *lloyka* (*Sturnella loyca*), which like the shaman or *machi* obtains its medicines from the forest plants— illustrate how profound are the notions of belonging to ecosystems and the identification of humans with birds. Moreover, the Yahgan and Mapuche narratives teach us about practices and values that impel this notion of pertaining to biotic communities.

The Yahgan story of the hummingbird, or omora, (*Sephanoides sephaniodes*), for example, has a clear ethical message. Food and water should be shared between all members of human and biotic communities, and with his act *omora* brings social as ecological well-being. This *omora* story shows how the Yahgan-bird community achieved the provision of drinking water, and teaches us that social well-being and conservation of biological species go hand in hand; that solidarity in the end is better than selfishness, and leads us toward a better life for all. This is especially true in the extreme south.

The perspective of conservation of bird and human communities in the Yahgan story of *omora* is comparable to current ecological statements. Modern sciences propose that birds and other species contribute to ecological processes, such as pollination, the dispersal of seeds, the control of insect populations, preying upon rodents and the cycling of nutrients with their guano and the mineral salts that birds transport from ocean to mountain ecosystems in the austral fjords. At the same time, the maintenance of these ecosystem processes contributes to the conservation of the flora in watersheds. This vegetation is crucial for the regulation of hydrologic flows, because it contributes as much to preventing floods as droughts. In this way, the Yahgan narrative of *omora*, as well as the scientific ecosystem analysis, recommends the protection of biodiversity of the watershed in order to secure the equitable and sustained provisioning of drinking water.

From the perspective of contemporary environmental ethics, the imperative implicit in the Yahgan narrative and the scientific analysis is to conserve the bird community in order to assure reliable water and, in this way, to respect the natural right to life of the diverse birds, human beings and other living beings. This imperative emphasizes the notion of *instrumental value,* because the conservation of biodiversity is an *instrument* for the survival of humans and the assemblage of other species.

The notions of *intrinsic* and *instrumental value* of the life of birds, humans and other living beings that co-inhabit the region of the temperate forests of South America, derive as much from current scientific analyses, as from traditional ecological knowledge. Their ethical-practical purpose is to promote the continuity, and continuance, of life.

An understanding of the diverse languages of cultures and birds, permits us to feel and look at ourselves immersed in the forests, living together with the birds and with the myriad of perceptions we humans can have of the birds. As human beings of different sexes, ages, geographies, cultures and histories, the austral forests open to us an opportunity —in the context of a globalized world— to glimpse and discover very diverse biocultural corners. This Multi-ethnic Guide is a source book of reflections about how to live together respectfully with the singular biocultural diversity that inhabits the extreme south of the Americas. At the same time, its voices and narratives can inspire respectful forms of inter-specific and inter-cultural inhabitation of the multiple urban, rural and remote corners of the planet, each one characterized by unique biocultural diversities.

IV. Songs and stories recorded in this guide

The song of each bird can vary from one moment to the next, between spring and winter, from individual to individual, between one forest and another—as occurs with the *chámuj*, *chinkol*, *chincol*, Ruffous-Collared Sparrow or *Zonotrichia capensis*, whose trills vary between night and day and from northern to southern populations.

Like the voices of the birds, the names of and stories about birds told by the Yahgans, the Mapuches and the scientists change from moment to moment, from place to place and from person to person. For example, *omora* is more hummingbird-human in the story recorded by Grandmother Julia; in contrast, it is more human-hummingbird in the narrative told by Grandmother Rosa, but always *omora* is small and courageous. So, we find variations and similarities among stories, even within the narrations of a given story. The accounts of this guidebook constitute only a moment in the scientific, Yahgan and Mapuche ornithological record. The search remains endlessly open.

In this guidebook we find particular records of birdcalls and stories that present many variations through the geography and history of southern South America. Let us get to know, then, something about those who tell the particular Yahgan and Mapuche narratives recorded in the CDs of this guidebook.

Brief Biographies of Úrsula Calderón and Cristina Calderón

Cristina Zárraga

Úrsula Calderón Harban was born on the 7th of September, 1925 in Mejillones Bay, Navarino Island. After the death of her mother, when she was barely 7 years old, she lived with her half-sister Dora and her brother Juan. When she was 14, she worked on the Róbalo Ranch, and at age 15, she married José González. Together, they worked hunting sea otters, and during eight years they lived moving from one place to another by boat. They resided in in their tent in Puerto Navarino or simply where they anchored at night. Later, they set up residence on Mascar Island, but, as their children needed to attend school, Úrsula moved to Puerto Williams with the children. José remained on Mascar, raising cattle and sheep. Later, he too lived in Puerto Williams, where he died in 1987, brought down by lung cancer. Today, Úrsula lives with her three children in her house in Villa Ukika in Puerto Williams, and she dedicates herself to handcrafted artistry – weaving baskets of rush, building canoes, needlepoint and knitting – in addition to speaking her native language.*

Grandmother Cristina Calderón (left) and her sister Úrsula in the house at Mejillones Bay, Navarino Island, during a break while recordings were made for this multi-ethnic guidebook in February 2001.

**Grandmother Úrsula Calderón died in January 2003. Today she rests in the Cemetery of Mejillones, Navarino Island.*

Cristina Calderón Harban was born on the 24th of May, 1928, in Róbalo Bay, Navarino Island. Her infancy passed between the bays of Róbalo and Mejillones. At a very early age her mother died in the latter place. As a result, she remained in the care of her grandfather, Alapainche, who in a short time also passed away. Then she lived with her cousin Clara and also her aunt Kerti. At the age of 16, she was required to live with Felipe Garay, a good man who, although quite a bit older than she (50 years), cared for her during the years that they lived together at Eugenia Ranch at the northern coast of Navarino Island, until the death of Felipe. From this union three children were born. At age 19, she met Luis Zárraga, a man descended from the disappeared Ona ethnic group. With him and the children, they lived at Harberton Ranch in Argentina for nine years, where six more children were born. Then, they moved to Puerto Williams on Navarino Island, building their house in Villa Ukika in 1960. Later, Luis became sick, dying later in Punta Arenas. In 1964, Cristina united with Teodocio González, with whom she shares her life until today. From this union one daughter was born. Cristina dedicates herself to handcrafted artistry – traditional weaving of rush baskets, making Evergreen Beech bark canoes, knitting and painting – in addition to fluently speaking the Yahgan language.

Villa Ukika, August 2001

Grandmother Cristina Calderón weawing a basket made of rush or ushkulampi (Marsippospermum grandiflorum) in her craft working place in Villa Ukika, Navarino Island, June 2001.

Short Biography of the Mapuche Bird-Man or Üñümche

Lorenzo Aillapan Cayuleo

He is born in the community of Rukatraru (habitat of the *traru*) on Lake Budi (*Furi leufü*), which lies back to back with the great Pacific Ocean on the southern cone of South America, and is the region's only saltwater lake. He is born with the name *Llankaleu* (precious stone or pearl of the water).

He is the son of farming parents and the grandson of the powerful *lonko* (chief) Aillapan (nine lions or pumas). The *lonko* Aillapan is and was a warrior poet of the Mapuche style before and after the so-called "Pacification of the Araucanía" (when Chile finally took control of the Araucanian or Mapuche territory in about 1880).

The child Llankaleo (u) later will be named Lorenzo by the Civil Registry due to the demands of the Chilean law that require "saint's names" for people. At the age of 8 to 9 Llankaleu is consecrated and initiated through a dream (*pneuma*) as a bird-man (*üñümche*), bird spirit of the ancient Mapuche cosmovision. This consecration was guided by renowned masters: Master Huaiquián, Master Imio, Master/nuke Kayuleu (six rivers). The university of life with the guidance of wise masters gave him the title of *üñümche* (bird person) as an inheritance for his whole life; the community and all the people approved the title of Mapuche Bird-Man – until today.

At age 11, Lorenzo went for the first time in primary school (the Methodist School of Rucatraru), and he began to understand the Chilean/Spanish language. After completing his primary education, with the help of the North American benefactors Rann Crawford, Husell Sargent and Nancy Sargent, Lorenzo went to the Methodist Farming School of Nueva Imperial. He also studied at the Men's High School of Nueva Imperial. Later, he completed his higher education and worked in Santiago, the capital of Chile.

Traytrako, August 2001

The poet Lorenzo Aillapan Cayuleo, Mapuche Bird-Man or Üñümche, with photographer Oliver Vogel (center), and ornithologist Steven McGehee (left) at the entrance of the Omora Ethnobotanical Park, Navarino Island, on February 27, 2001, after completing the recordings for this multi-ethnic bird guide.

Acknowledgements

As was said in the Introduction, this guide is a collective work that is possible thanks to the fact that birds, forested ecosystems and very diverse people exist. In particular, the authors sincerely thank the following individuals and organizations:

Mario Chiguay, President of the Indigenous Yahgan Community of Bahía Mejillones, and all the other members of the community who collaborated in this project, especially Julia González;

Manuel Muñoz, Advisor to the Chiloé Council of Chiefs, the members of the Huilliche communities in Chanquin and Huentemó, the school teachers of Chanquin and Mr. Francisco Delgado, Chiloé National Park, who encourage and helped the initiation of this project;

the poet Lorenzo Aillapan, who teaches in the Pullümapukimunweftuy Mapuche Academy in Puerto Saavedra, a cultural center which this guide hopes to serve;

Dr. Víctor Fajardo, President of the University of Magallanes, Luis Oval, Vice-President for Academic Affairs, and the professors of the Department of Natural Science Orlando Dollenz, Sylvia Oyarzún and Dr. Andrés Mansilla; this guide book also forms part of the biocultural and environmental ethics program, initiated by the University of Magallanes and the Omora NGO at the Omora Ethnobotanical Park in the Chilean Antarctic Province;

the Chilean Antarctic Provincial Government and Governor Eduardo Barros, who actively participates in the search for equity and conservation at the southern tip of the world, and graciously received poet Lorenzo Aillapan into the provincial government's guesthouse and supported the work of this guide;

the National Natural History Museum, which provided access to its valuable collection that demonstrates Chile's biological and cultural diversity; we especially acknowledge the cooperation of the museum's director, Eliana Ramírez, and Curator of Ornithology Juan Carlos Torres-Mura, who has been key in completing this project;

the Center for Biodiversity and Conservation at the Department of Ecology and Evolutionary Biology – University of Connecticut (USA), which permitted the necessary research for the preparation of distribution maps, as well as the manuscript of this book; Ricardo Rozzi profoundly thanks the members of his doctoral committee Dr. John Silander, Dr. Gregory Anderson, Dr. Robin Chazdon, Dr. Robert Colwell and Dr. Steward Pickett, as well as Dr. Robert Dewar and Dr. Christopher Elphick, at the University of Connecticut for their ecological, anthropological and ornithological guidance during the development of this guide book;

the cyber and personal exchanges by way of the Network of *Enseñanza de la Ecología en el Patio Escolar* (*EEPE*), which remains a fount of learning and constant stimulation; particular thanks to the network's coordinators Dr. Alejandro Grajal and Dr. Ricardo Stanoss of the Audubon Society's Latin America and Caribbean Program (USA), we hope that this guide will be an educational source material to aide in the search for ways to live respectfully with the world's biological and cultural diversity by showing the urban, rural and remote corners of Latin America;

Leonard Yannielli and the volunteers from the Earthwatch Institute project *The Owls of Cape Horn* for their collaboration in the recording of birdcalls and the study of the natural history of the avifauna in the austral extreme of the Americas;

the Senda Darwin Foundation, especially Dr. Juan Armesto and Dr. Mary Willson; Claudia Hernández, Claudia Papic, Marco Méndez, Emma Elgueta and Iván Díaz, who cooperated in the ornithological research; Hernán Rivera from the Chilean National Parks Service and Guillermo Egli, of the Union of Chilean Ornithologists, with whom we organized the workshop *Recognizing the Forest Birds of Chiloé*, conducted in Chiloé National Park in December 1995. In this workshop participated: Francisco Delgado, José Gallegos, Mario Guinao, Jeno Muñoz, Jorge Panichini, Roberto Rosas, Jose Subiabre and Carlos Uribe (Chilean national parks agency), Eduardo Ramilo, Juan Salguero and Lorenzo Simpson (Argentine parks delegation), Enrique Couve, Luis Espinoza and Andreas von Meyer (Union of Chilean Ornithologists), Carlos Sabag (Senda Darwin Foundation), and Andrea Bachmann and Lily Sheeline (Peace Corps – USA); all helped precipitate the initiation of this book;

the recording studios of Button Records (Puerto Williams) and Radio WHUS-University of Connecticut (USA), which offered their facilities and studio time, under the direction of Nolberto González and John Schwenk respectively, for the editing of sound tracks that allowed the combination of bird calls with stories and the voices of different people in the CDs of this guide; without this technical and infrastructure cooperation for the sound archives, this guide would not have been possible;

the composers and students of Taller Matta 365, Valparaíso, who have given new life to the multiple bird and human voices in their experimental music workshop; Andrés Alcalde for his music and for the piece for piccolo *Gli Uccellini*, played by Guillermo Lavado;

Patricio Riquelme for his friendship, his art and the design of the CD cover; artists Daniel Martínez and Jay Barry for their drawings (see Photograph and Figure Credits) and all those who donated their photographs (see Photograph and Figure Credits);

Enrique Couve and Claudio Vidal, Fantástico Sur Publishing, for welcoming us in the alliance between science, educational tourism and birds;

Dr. Mary Kalin-Arroyo, Millennium Center for Advanced Studies in Ecology and Biodiversity Research - University of Chile, and Jorge Pavez, America's Fund, for constant encouragement and support; the projects *P99 103 FICM* and *Biological and Cultural Conservation in the Austral Extreme of America* permitted the co-financing of this project in conjunction with the University of Magallanes and Fantástico Sur Publishing;

Humberto Maturana, Baird Callicot and Eugene Hargrove for their works on experimental epistemology, comparative environmental ethics and the intrinsic value of all living beings that inspire this guide;

the members of the Omora NGO, especially Héctor Massardo, Andrés Marin, Maurice van de Maele, Livia Firmani, Bryan Connolly, Augustin Berghöfer, Margaret Sherriffs, Alicia Lavanchy and Alejandro Correa, and to all the persons who made this education project possible.

EXPLANATION OF THE LAYOUT, AND SYMBOLS USED IN THIS GUIDE

Sequence of birds. This guidebook includes the names and narratives of fifty bird species. The sequence of birds follows an imaginary journey from the forest interior towards the margins and adjacent open habitats. The order of this succession of species is the same in the illustrated text and on the CD name tracks.

The sequence includes:

(1) **birds that inhabit either exclusively or principally the forest interior** (all these species are endemic to the temperate forests of austral South America);

(2) **owls that can be observed in the forest interior** (as with the birds in the previous group, these are principally detected by their call);

(3) **birds associated with riparian, coastal or bog habitats** (when one wanders through the austral forests it is frequent to cross or find courses of water, lakes, fjords, channels, rush thickets and other wetland habitats);

(4) **birds that are observed exclusively or principally along the border of forests and/or in adjacent, open habitats** (these species are easier to detect visually given the fact that they are found in open places, in addition to usually being more generalist in their habitat use);

(5) **raptors that usually are seen perched on tall trees or soaring over the austral forests or adjacent types of habitat** (raptors fly at great altitude over the forests and a great diversity of other habitat types; they daily travel greater distances than the other types of birds and their geographic ranges tend to be extensive, as well) .

Bird names. The description of each bird species begins with its Yahgan, Mapudungun, Spanish, English, and scientific names. This order is repeated throughout the text and CD recordings. In the recordings it was not deemed necessary to record the scientific names.

In the case of the recordings only the most used names were taped. For Yahgan and Mapuche names we only recorded those known or preferred by the Yahgan grandmothers Úrsula and Cristina Calderón, and by the Mapuche poet Lorenzo Aillapan, respectively. These names appear in the text with bold letters and are at the beginning of each language name list. Inlighter font other names are also included that came from other ethnographic interviews or bibliographic sources. In a similar way, the common names in Spanish and English that are recorded on the CDs are put into bold letters in their respective lines in the text, but are followed by other, less common names in lighter font.

As you may have already noticed in some passages of the Introduction, some names and other terms are written in color:
blue is Yahgan;
red is Mapudungun;
and green is used for the common names (including other species of flora and fauna) in English.

Regarding the writing of the Mapuche language, these were written in conjunction with Mr. Aillapan, who closely follows the Unified Mapuche Alphabet, proposed by linguist María Catrileo. The Yahgan terms recorded with the grandmothers Úrsula and Cristina Calderón were transcribed as closely as possible to an equivalent in Spanish pronunciation. Using the recordings of this guide, these names are being transcribed by linguists to the International Phonetic Alphabet (IPA). The Yahgan and Mapuche names that are taken from bibliographic sources maintain the original format of the source.

Distribution maps. The description of each bird species includes a map showing its geographic range, taking into account season movements and/or degrees of abundance. These maps use the following color code:

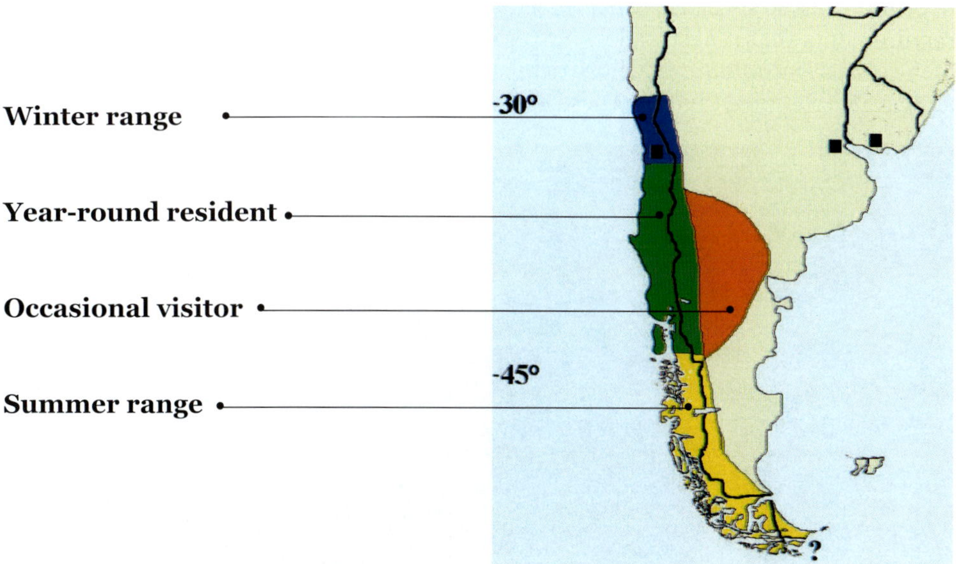

Texts of the narratives recorded on the CDs. The symbol indicates the texts of the narratives recorded onto CDs. For the Mapuche stories, the Mapudungun texts were transcribed together with the Spanish texts.

Yahgan: **Lána**, Lana, Kanára
Mapudungun: **Kurüpütriu**, Concona, Concoma, Rere, Pitigüe
Spanish: **Carpintero negro, Carpintero gigante**
English: **Magellanic Woodpecker**
Scientific: *Campephilus magellanicus*

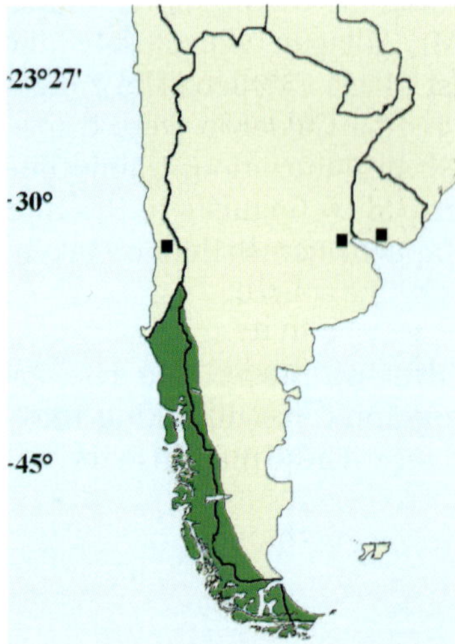

The Magellanic Woodpecker is the largest woodpecker species in South America and, probably, the world. It belongs to the same genus (*Campephilus*) as the two largest known species of woodpeckers, the Imperial Woodpecker (*C. imperialis*) and the Ivory-Billed Woodpecker (*C. principalis*). Both species inhabited forests of the Northern Hemisphere and are presumed to be extinct today. The Magellanic Woodpecker is endemic to the *Nothofagus* forests of southern Chile and Argentina. It is so specialized in its habitat requirements that it nests solely in trees of the genus *Nothofagus* in whose trunks it excavates rounded cavities. These holes

provide nesting sites not only for woodpecker families, but also for numerous other cavity-nesting birds, such as the Austral Parakeet (*Enicognathus ferrugineus*) and the Austral Pygmy Owl (*Glaucidum nanum*). This woodpecker creates homes for a variety of birds and is, therefore, a keystone species in the austral forest ecosystems.

Magellanic Woodpeckers also require old trees to obtain the insect larvae it digs out from rotten wood with its strong beak and tongue. To find appropriate trees to feed upon and live in, this woodpecker requires large territories, in addition, it nests mainly in the interior of old-growth forests, more than 100 m away from forest edges. For these reasons, Magellanic Woodpeckers are very sensitive to the degradation of the austral temperate forests and their replacement by exotic plantations of *Eucalyptus* and pine (*Pinus radiata*). Therefore, they require special conservation measures if they are to avoid becoming extinct like their congeneric species from the Northern Hemisphere temperate forests. Fortunately, it is not too late to save the Magellanic Woodpecker. Although it is a severely endangered species in Chile, it is still possible to observe males moving around with their bright-red, long crests and to listen to woodpecker families calling and pecking in the dense and remote forests that are still intact.

In territories that preserve large expanses of austral forests there will almost always be small families of woodpeckers marking their presence with the loud drumming-noise of their pecking. On Chiloé Island, where the woodpecker is known as *rere* or "rooster of the mount," their "wood cutter's" drumming echoes throughout the mountains. This intense sound is associated with the chopping of the mythic *trakco's** hatchet. The Magellanic Woodpecker also receives the Mapudungun name of *kurüpütriu*, which describes it as a black (*kurü*) carpenter (*pütriu*).

The large Magellanic Woodpecker or *rere* reaches the most southerly distribution of all the woodpeckers in the world, inhabiting even the forests of the Cape Horn Archipelago, where the Yahgan know it as *lana*. The name *lana* derives from the Yahgan word *lan* that means tongue, referring to the long and strong tongue that the woodpecker uses to extract the larvae. The Yahgan grandmother Úrsula Calderón remembers that when she was a little girl, she very much admired the skill with which the woodpeckers, or *lana,* pecked holes and extracted grubs or white larvae with their tongues, which they swallowed throwing their heads back. Her father, Juan Calderón, told her that woodpeckers were descended from humans.

Yahgan story. The Magellanic Woodpecker or lana in the forests of the Cape Horn Archipelago accompanied the Yahgan women when collecting dihueñes in the forests. A Yahgan grandfather, Juan Calderón, said the origin of this beautiful bird of the austral forests goes back to ancestral times when birds were still humans.

In those remote times, a brother fell in love with his own sister and tried every kind of trickery to be with her and sleep together. His sister noticed this intention and eluded her brother each time to avoid taboo relations, but in the end she was of two minds: half of her wanted her brother and half of her did not.

The brother, who continued thinking of pretexts to bring her away from the akar, or hut, discovered one day giant, red chaura or Prickly Heath (amai, Gaultheria mucronata) berries in a forest gap and quickly went to tell his sister, "I have found enormous chauras in a place in the forest. You should go there and collect them." The sister took her basket and went into the forest, while her brother followed without anyone noticing and hid himself in a place where his sister had to pass. When she came sufficiently close, he embraced her, and they fell to the ground giving course to their love.

When they stood up, they converted into birds and flew like lana. Since that time, they live together in the forests, and the brother carries over his head a red crest that reminds one of the color of those big chaura fruits.

* *The trauco is a small mythic inhabitant of the forests of Chiloé Island. He has a human figure and dresses with branches of the vine* Luzuriaga radicans. *He carries a stone hatchet that he uses to gather fruits and vegetables, and glean food from rotten trunks. He walks through the forests pecking with the hatchet, like a rere.*

Yahgan: **Túto,** Tútu
Mapudungun: **Tiftifken**, Churin, Chercan, Chircan, Chochif
Spanish: **Churrín,** Churrín andino, Pa' atrás
English: **Magellanic Tapaculo,** Andean Tapaculo
Scientific: *Scytalopus magellanicus*

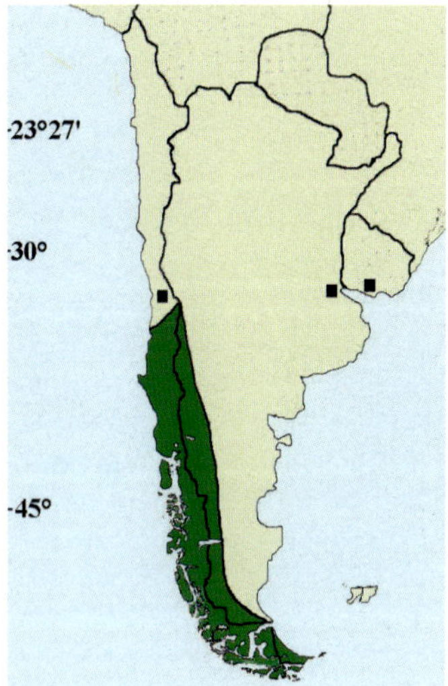

From far away, the tiny Andean or Magellanic Tapaculo is heard when it emits its intense territorial call *"churin, churin, churin."* The *Mapudungun* name *churin* echoes the call, and has been adopted by Spanish-speaking Chileans. As the *Lafkenche* poet Lorenzo Aillapan explains, "because of the regularity of the *churin's* vocalization it calls to mind a pocket-watch." Therefore, the *Lafkenche* have metaphorically named this little bird the *tiftifken* or "pocket-watch bird".

The Magellanic Tapaculo is dark and difficult to see because, like the Chucao Tapaculo and the Black Throated Huet-Huet, it lives immersed in the dense understory vegetation of the austral forests, preferably near streams of water. However, in contrast to its relatives, it sometimes ventures forth to the margins of the forest and even is found in dense brush habitat, especially on the islands in the southern channels. Perhaps because of its ability to live in diverse habitats, the Magellanic Tapaculo is the only Rhynocriptidae to reach the world's southernmost forests. Although theTapaculo family is tropical in origin, the Magellanic Tapaculo was first described based on a specimen collected in Tierra del Fuego in the 18th century; for this reason, *magellanicus* is its species name.

The Magellanic Tapaculo feeds on insects while it moves among the lower branches and the ground. In the understory, in trunk cavities or between large branches, it carefully builds its nest with fibers, roots, brackens, lichens, and mosses. Its unique ability to build these nests contrasts with the other tapaculo species of the austral forests. Perhaps, ths skill of the Magellanic Tapaculo inspires the Yahgan story that honors this bird, túto, with its terrestrial lifestyle and poor flying capacity, for inventing the *anans*, or canoes. These canoes were made of *shapea* bark, or Magellan Evergreen Beech (*Nothofagus betuloides*), and allowed the Yahgans to navigate through the channels, bays and fjords of the Cape Horn Region.

Yahgan story. An ancient Yahgan story tells that in the time of the ancestors, when birds were still humans, the Magellanic Tapaculo or *túto* was an old widower. He lived a long time with his son-in-law, who received the old man into his *akar* or hut. As the old man did not have other relatives, his son-in-law cared for him and maintained him during the last days of his life. However, the old man was always unsatisfied, even though the son-in-law took great pains, and every day went to the beach to collect him shellfish, meat and, occasionally, fruits from the forest. Nothing was to the old man's pleasure, and although he ate everything his son-in-law brought him, he always said, "I am already very old. I'd like to chew on something that I really like." So, the son-in-law looked for other things and brought something new each day, until one day he brought upon his return to the *akar* strips of *uri**, which the father-in-law chewed with great pleasure. He did not complain, but rather with great satisfaction said, "This is what I like."

When the son-in-law heard these words he sighed and said very happily, "At last, how happy and complacent my father-in-law is with the fibers of *uri* that I brought him. Now I will bring him many fibers. How could I have known that he would like them so much?" He went each day and always brought a bundle of *uri* that he put before his father-in-law. *Tuto* began to chew the fibers immediately, and after having carefully chewed a great quantity of fibers, he took pieces of bark and squeezed them with the fibers of *uri*, thereby inventing the *anans* or canoes that the Yahgan used to travel through the austral archipelagos. The old man made many canoes and gave them to the people. Since that time, one always needs *uri* to stick together the pieces of bark, so the canoes withstand the waves for a long time.

* Uri *is the Yahgan word for the fibers that can be extracted under the bark of the Austral Beech (*Nothofagus antarctica*).*

Yahgan: out of bird´s range
Mapudungun: **Wed-wed**, Hued-hued, Huahueta, Huid-huid
Spanish: **Huet-huet,** Hued-hued
English: **Black-Throated Huet-Huet**
Scientific: *Pteroptochos tarnii*

The Black-Throated Huet-Huet is the largest Passerine bird that inhabits the temperate forests of austral South America. It lives in the interior of dense jungles between the channel forests in the area of Wellington Island (49° S) to the Bío-Bío River (37° S). North of the Bío-Bío River until the Tinguiririca River (35° S) lives its sister species, the Chestnut-Breasted Huet-Huet (*Pteroptochos castaneus*). Hidden among the understory, the Huet-Huet makes short flights and frequently is found scratching the soil in search of food that includes a great variety of invertebrates, seeds and fallen fruit. They also excavate large galleries in the earth that serve as nests and sleeping sites. Dark colored, this Rhynocriptidae is difficult to see among the dense vegetation.

Despite its ability to blend visually into its habitat, the presence of the Huet-Huet is signaled by its intense calls that arise from the bottom of ravines or thickets of the southern rainforests and are audible from great distances—especially its territorial call "*woo-woo-woo,*" which is characterized by a series of descending notes. Its repeated warning call *wed-wed, wed-wed* gives rise to its onomatopoeic *Mapuche* name *wed-wed*, from which is derived its Spanish name *huet-huet*. In *Mapudungun*, *wed-wed* generally means "crazy" and, more specifically, a person with a joke flowering on his lips who knows that what he says is funny and dares to say, and show, it. The *wed-wed* sounds like the talking of a crazy person who, from far away in the ravine, accompanies travelers through the forests, narrating curious stories for their pleasure as they go.

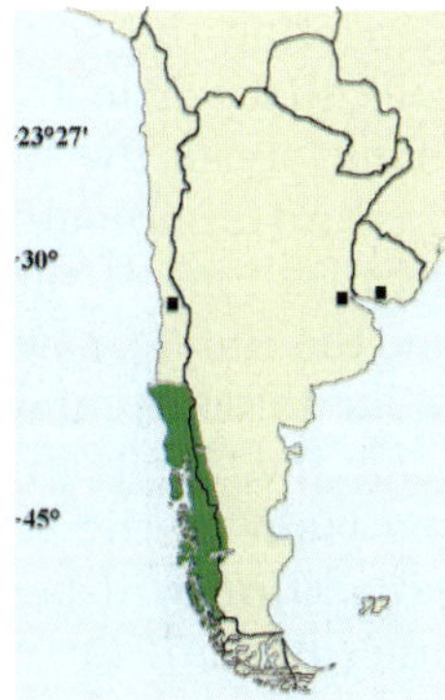

Additionally, this terrestrial bird has many varied songs that the *Huilliche* interpret on Chiloé Island in order to forecast the weather. The call "*dehuet-dehuet*" predicts rain; the call "*mahuín-mahuín*" announces drizzle; when one hears the penetrating "*uh, uh, uh*" coming from the ravines, there will be sun.

Yahgan: out of bird´s range
Mapudungun: **Chukau**,Chucao, Chucau, Tricau, Chiduco, Chukaw
Spanish: **Chucao**
English: **Chucao Tapaculo**
Scientific: *Scelorchilus rubecula*

Repeating the sound of the *chukau's* calls, that peal forth from the understory of the forests, the *Mapudungun* name of this bird is onomatopoeic. However, the vocalizations of the *chukau* are varied and the *Huilliche* hear distinct messages in its different modulations, when they roam the forests. If the *chukau* sings to the right of the forest wayfarer, the expedition will be beneficial. On the other hand, if the call of the *chukau* is heard from the left, the fortune will be adverse. If the call of the *chukau* is constant and penetrating, then this announces rains and storms.

The Chucao Tapaculo, like the Black-Throated Huet-Huet, is an endemic species that characterizes the diversity of the temperate forests of southern South America. It also lives in forest interiors with dense canopy and a well-developed understory, where it is one of the most common birds. While moving among low branches or walking on the forest floor, the *chukau* forages for invertebrates and fruits fallen from trees, sometimes excavating the earth in its pursuit. This terrestrial bird collects twigs, herbs, and lichens to build its nests in the holes of trees or hillsides.

Chucao males are very territorial and stay in the same place for six years or more. They defend their territories by emitting loud calls, and by physically bumping chests with their competitors. Their short, loud, sharp and manifold vocalizations epitomize, aurally, the character of the forests of southern Chile and Argentina.

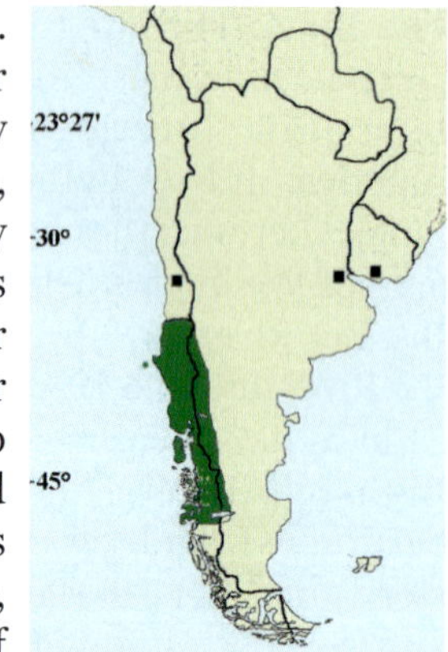

The Chucao Tapaculo is a beautiful bird with a bright orange throat and chest, yet it is difficult to see because it lives hidden among branches. The *Lafkenche* tell that the *chukau* has the habit of walking

secretly, bent over in the understory of the vegetation because it is like a beautiful maiden who was surprised by a young man when bathing under a waterfall or *trayenko* (*ko* = water, *trayen* = the naturally romantic sound of water).

Mapuche story. In Lafkenche territory, lovers court in the thicket of the forests, where many romances begin. The maidens go to the forest, also, to look for water or to take baths in the streams and waterfalls. It is told that the *chukau* bird has the habit of walking hidden in the thickness of the vegetation because long ago she was a beautiful maiden who was bathing under a waterfall or *trayenko*. She was surprised by a young man and felt embarrassed. Since that time this bird with its little tail standing up and its little legs towards the front wanders through the forests, like a maiden showing herself timidly, and at the same time coquettishly, between the foliage, where she emits her cries and call. In the old *Huilliche* weddings on Chiloé Island, which began with the abduction of the bride, the couple returned from the forest only once the *chukau* wished them all good luck with its call.

Mapudungun version. *Pu Lafkenche püle, üñawen peukey pu mawidantu mew, cheutañi müleken epuñpü e poyeuwün. Tachi pu malen komekeg rangintu mauwida meu müñetuaal mangiñko ka trayenko meu. Feyti weda chukau may ellkau miyaukey rangiñ karii lelfan ka fey witraley ñi pichi külen ka ñi epu namun rulalkuley, fey mag feyti allangechi we malen femngey müñetumekelu trayenko reke meu fey may perkey kiñe domokulmen wecheweutru. Fey may rumel müyaukey ñi am feyti malen lluru trekayawi rangiñ mawidantu pichike llükaley, welu duan niefuy chem, rangiñ tapül mawida, fey wefküley ñi mütrüm ül. Kayfimel pu kurewen williche CHÜLLEWE (Chiloé), feychi meu weñenge kefuy we malen, kurewen wüñotukefuy mawida meu feyti CHUKAU, Küme üL ülkantu wepum rumel.*

Yahgan: **Tatajurj,** Tatajúrj
Mapudungun: **Pelchokiñ,** Pishonquillu
Spanish: **Comesebo grande**, Comesebo, Picolezna, Carpintero carmelito
English: **White-Throated Treerunner**
Scientific: *Pygarrhichas albogularis*

The White-Throated Treerunner is endemic to the temperate forests of southern South America, and it is the sole representative of the genus *Pygarrhichas*. It lives in the forest interior, where it can be observed picking trunks in the mid-level and high canopy like a little woodpecker.

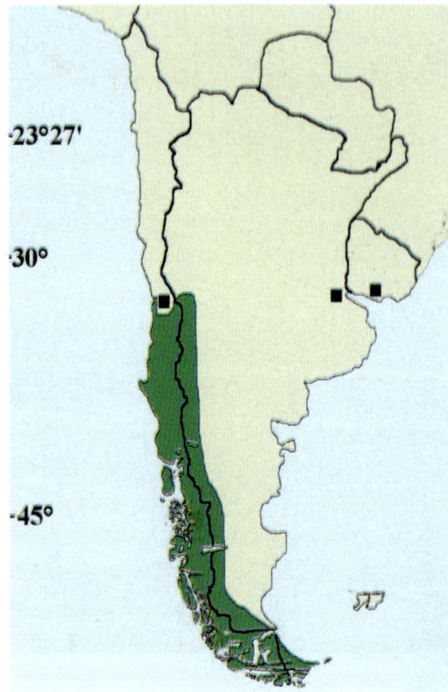

The White-Throated Treerunner flits up and down the trunks and large branches of old trees, vigorously boring its own holes and feeding on the small insects and larvae that it finds. Its behavior shows us how important it is to conserve the old and decaying trees that provide the basis of existence for it, and many other living beings, that like the White-Throated Treerunner are scarce and habitat specialists.

Although the White-Throated Treerunner is not large, it is conspicuous due to its prominent white throat and breast that inspired its scientific name *albogularis*. To hear this bird, you must pay attention to the short and low metallic sounds that resemble a rapid succession of drops. Outside the breeding season, they move around in forested areas, frequently forming mixed flocks with the Thorn-Tailed Rayadito (*Aphrastura spinicauda*).

This bird reaches the world's southernmost forested watershed on Horn Island (56°S), and traditional Yahgan stories mention the White-Throated Treerunner, or *tatajurj*, accompanying the Yahgan women or *kipa*, when they collected *katran*. Like the White-Throated Treerunner, these fungi of the genus *Cyttaria* are endemic to *Nothofagus* forests. The Yahgan grandmother Cristina Calderón stills remember how much she enjoyed watching the *tatajurj* creeping up and down along the tree trunks.

Yahgan story. When I was a girl in Robalo, I always liked to go into the forest looking for twigs, firewood for grandfather Wisch and grandmother Hamumo, and in the forest I always saw the *tatajurj*. I liked them a lot. Sometimes I sat down for a long while to watch them, how they went up and down again and again, pecking the trunk. There were always two of them going together, the couple of *tatajurj*. I liked them so much that I was late in bringing the grandparents the firewood, and when I came back from the walk into the forest, they asked me, "Why are you so late?" And I told them that I did not find any firewood. Therefore, I had come late. I didn't tell them that I was watching these little birds.

**Grandfather Wisch (or Whaits) told several traditional Yahgan stories to the anthropologist Martin Gusinde, who wrote the largest Yahgan ethnographic record (see References). According to the Yahgan custom, the children took care of old people bringing them water and fire-wood. Today, Robalo Bay is geographically a part of the Omora Ethnobotanical Park. This Park is an institution that conducts ethnoecological research and education programs, including the preparation of educational materials such as this Multiethnic Bird-Guide.*

Yahgan: **Tachikáchina,** Tečkášina, Tulératécix
Mapudungun: **Pidpidwiriñ,** Pidpid, Pishpish, Yiquiyiqui
Spanish: **Rayadito,** Raspatortillas, Yiquiyiqui
English: **Thorn-Tailed Rayadito**
Scientific: *Aphrastura spinicauda*

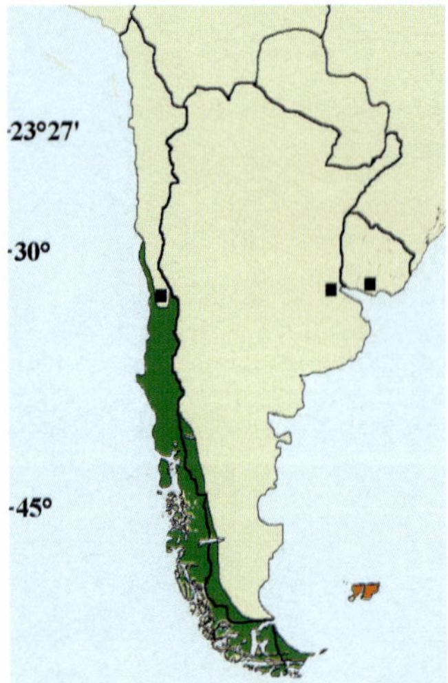

Aphrastura is another genus endemic to the forests of Chile and Argentina. This genus possesses two species: the Masafuera Island Rayadito (*A. masafuera*) which lives in the forests of the Juan Fernández Archipelago and is very endangered due to habitat destruction and introduced exotic predators; and the Thorn-Tailed Rayadito (*A. spinicauda*) which is found along the whole range of the *Nothofagus* forests, from Cape Horn to the relict coastal forests of Fray Jorge (31°S).

The Thorn-Tailed Rayadito is a little bird with white throat and breast, and yellowish-brown back plumage. Its common Spanish name *rayadito* (little striped one) is derived from the black lines that cross its crown, forehead, and wings. Its English (Thorn-Tailed) and scientific *(spinicauda)* names come from its prominent tail, which has protruding outer feathers with terminal points.

Thorn-Tailed Rayaditos move tirelessly among branches of the canopy and understory and on trunks, where they peck the bark, looking for larval and adult insects. They nest in holes of live or dead trees and stumps, and are able to pass through very narrow openings to enter to their nests. The nest provides protection from predators and contributes to their success in *Nothofagus* forests, where they are one of the most abundant bird species.

Thorn-Tailed Rayaditos are one of the most easily detected and seen birds in the interior of austral forests. They are very curious, and when you enter the forests they come close and emit constant noisy vocalizations that give them their onomatopoeic Yahgan name *tachikáchina*. The Yahgans appreciate *tachikáchina* because they announce the presence of any other animal or person. As Grandmother Úrsula Calderón tells us, the Thorn-Tailed Rayadito is a watchguard that detects and announces dangers.

Yahgan story. *Tachikáchina*, or the Thorn-Tailed Rayadito, is a bird that sings during the day in the mountains to warn that somebody is hiding, a bad man, a witch-doctor. He tells the one who comes walking about the presence of others, also about any dog, cat.... well, about any hidden thing. Their cries make you afraid when they sing together, *tsch, tsch, tsch*, as they alert; nothing good.

Yahgan: **Kinan,** Kĩnan, Kínan
Mapudungun: **Choroi,** Kachaña, Rawikma, Cachuña, Tricau, Rawilma, Chilkeñ
Spanish: **Cachaña**, Catita austral
English: **Austral Parakeet,** Chilean Green Parrot
Scientific: *Enicognathus ferrugineus*

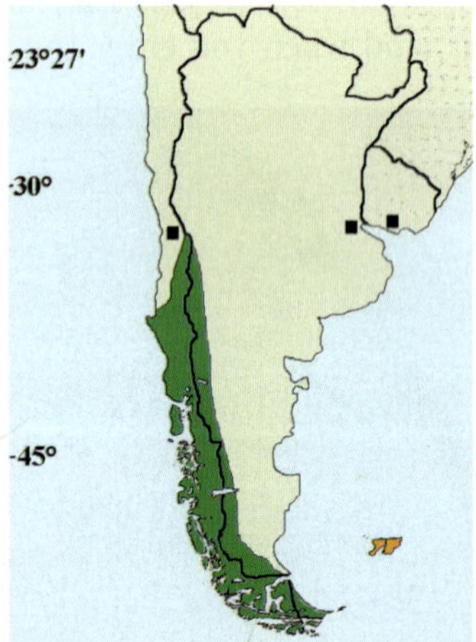

The parrot family has a near-worldwide distribution, but the majority of the species are concentrated in tropical latitudes. For that reason, when the French naturalist Bougainville referred to parrots that lived in Tierra del Fuego, scientists of the 18th century did not believe him. Later, the expedition of HMS Beagle and the naturalist Charles Darwin confirmed that the Austral Parakeet is indeed an exceptional parrot, which reaches the southernmost forests of the planet, inhabiting even the islands of Cape Horn. Moreover, the Austral Parakeet remains there year-round. Lucas, son of the Anglican missionary Thomas Bridges wrote, "these birds do not migrate, and in winter, although it might seem misplaced, the parrots can be seen in the canopy of trees covered by snow."

Bridges also wrote that the Yahgan and the Selknam, or Ona, Indians did not like the Austral Parakeet because they felt deceived by these noisy birds, which gave the guanaco hunters away to their prey as they followed them through the forest. For this reason, the Austral Parakeet was believed to be a spy, working for the guanacos. This does not surprise scientists, because a flock of austral parakeets usually has a sentinel bird that alerts its companions to the presence of predators or other danger with loud alarm cries.

Covering the entire range of *Nothofagus* forests in southern Chile and Argentina, gaudy flocks of Austral Parakeets fly over, or perch in, the canopies of Evergreen Beech (*Nothofagus betuloides*),Tall Deciduous Beech (*N. pumilio*) and other trees. They eat seeds voraciously, and nest in holes inside old tree trunks. Between the leaves and branches of *Nothofagus* trees, one can observe their showing bright green-yellowish plumage with rust colored tails, which is where its scientific name, *ferrugineus*, comes from. Austral Parakeets fly very skillfully through the canopy. They avoid obstacles and hunters. Denoting this ability, the *Mapudungun* name *kachaña* means "to elude." Furthermore, in Chile people use

the term "*cachaña*" (the Chilean name for the Austral Parakeet) to describe a person who eludes others with rapid and elegant movements.

Austral Parakeets are also admired for their capacity to imitate and speak. On Navarino Island, the Yahgan grandmother Cristina Calderón remembers that when she was a child and lived in Robalo Bay, there was a *kinan* (Austral Parakeet) that called her Uncle Juan by his name. Later, another *kinan* lived with grandmother Candelaria in Villa Ukika, where the Yahgan community lives today. That bird had the ability to imitate the cackling of the hens. In Antilhue, near the city of Valdivia, people feed captive Austral Parakeets with bread soaked in wine to teach them bad words. In country houses, it is common to find captive parakeets. If they are treated kindly, Austral Parakeets like the families who befriend them—but demonstrate a great antipathy if they are treated poorly. On Navarino Island, Grandmother Cristina prefers not to have captive *kinan*, however, because they suffer in cages and prefer to fly without hindrance. "Birds enjoy freedom," she says, "as much as we human beings do."

In the forests further to the north lives another parrot species, the Slender-Billed Parakeet (*Enicognathus leptorhynchus*). It is much less common than the Austral

Parakeet, and might well be called the "Chilean Parakeet" since it is found exclusively west of the Andes Cordillera.

The Slender-Bill is similar to the Austral Parakeet, but can be distinguished by its much longer bill, larger body size (40 cm), and its red forehead and lore. Its range of distribution is also much more restricted. The Slender-billed Parakeet lives mainly between the Bío-Bío River (38°S) and Chiloé Island (42°S), although it can be found occasionally between Valparaíso and Aysén. In the mountains of Temuco in the Allipen Valley is the town of Choroico; its name means water (*ko*) of the *choroi*. On Chiloé Island the Choroihue River means place (*we* or *hue*) of the *choroi*. This ornithological toponymy expresses how abundant this parrot species throughout the Mapuche territory. However, their populations have declined due to habitat destruction, hunting and because, like the Chilean Pigeon, it was affected by a fatal disease produced by the Newcastle virus in the middle of the 20th century.

The *choroi* especially typifies the Monkey-puzzle Tree (*Araucaria araucana*) or *pehuen* forests, where it feeds on the cones that it opens without difficulty, using its sharp, hooked bill. *Pehuenche* elders of the Andean zones of Galletue and Icalma still remember when, before they were hunted with rifles, *choroi* were so abundant that they clouded the sky. In March of each year hundreds of birds landed on *pehuen* trees to eat almost all their fruits.

According to the *Lafkenche* poet Lorenzo Aillapan, flocks of *choroi* were so numerous and voracious that they caused farmers to erect scarecrows. When flocks of this parrot landed on corn or wheat farms, they left nothing for the harvests. Consequently, *choroi* were intensively persecuted and hunted, contributing to their severe population decline. Today, both the Monkey-puzzle Tree and the Slender-Billed Parakeet are endangered, endemic species of Chile. We hope that acquaintance with and understanding about the history of the *pehuen* and the *choroi* can stimulate people to appreciating and preserving these ancient trees and birds, as well as the ecosystems and cultures with which they coexist.

Yahgan: out of bird´s range
Mapudungun: **Epukuden, Futrakülen,** Changkülen, Changkihue
Spanish: **Colilarga**, Cola de paja, Memorioso
English: **Desmur´s Wiretail,** Rusty Fern Tail
Scientific: *Sylviorthorhynchus desmursii*

Desmur's Wiretail is the only species of the genus *Sylviorthorhynchus,* and it is endemic to the southern temperate forests of Chile and Argentina. Thanks to its long tail, which can measure twice the bird's body length, it is an eye-catching species. There is no more amazing experience than to see it fly between the branches with its swaying tail, especially if it flies from one thicket of *küla* or bamboo (*Chusquea sp.*) to another. This bird is difficult to observe, however, given that it lives hidden in the thickness of the understory and its colors are not very attention grabbing, being yellowish-brown on the under parts and reddish-brown on top. Composed of *küla* roots or leaves, its nest is easily recognized by its spherical form with a side entrance.

The Wiretail feeds on insects that it hunts among the branches of the *küla* thickets. Its presence is discovered by its call, formed by reiterated *ye-ye-ye-ye*, or by its prolonged territorial trill that emerges from the brush similar to the song of the Tufted Tit-Tyrant (*Anairetes parulus).*

Its *Mapudungun* name *futrakülen* refers to the bird's long (*futra*) tail feathers (*külen*). On Chiloé Island the tradition exits of keeping a *futrakülen* tail feather in books to memorize them and understand better. For this reason, the *futrakülen* or "long-tail" is known in the Chiloé Archipelago as "the remembers-all" and students carry the tail feathers in their textbooks. They even sleep with these books underneath their pillows to aide in memorizing and as such to learn the lesson better. The other *Mapudungun* name *epukuden* might be related to this belief, since it means to foretell twice (*kuden* = foretell, *epu* = two).

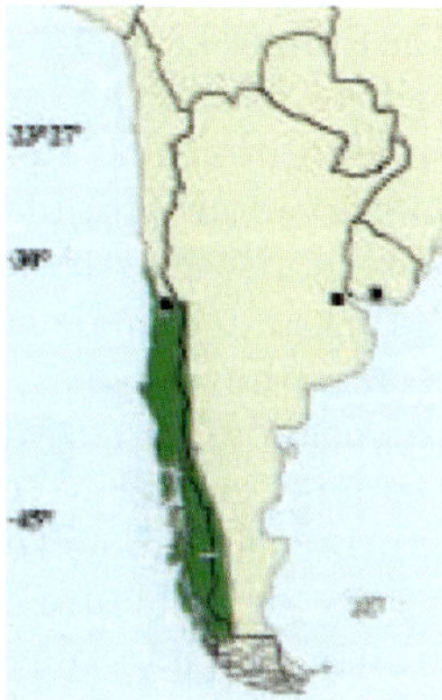

Yahgan: **Kuhúrj**
Mapudungun: **Kong kong,** Concon, Coa, Ñeque, Colcón
Spanish: **Concón**, Lechuza bataraz
English: **Rufous-Legged Owl**
Scientific: *Strix rufipes*

The Rufous-Legged Owl is a close relative of the Northern Spotted Owl (*Strix occidentalis*) that helped motivate the conservation of the old-growth forests of the Pacific Northwest of the United States and Canada. In the Northern Hemisphere or the Southern Hemisphere, these owls prefer to inhabit primary temperate forests with dense canopy cover. From the complex vegetation structure of the austral old-growth forests, sometimes incorrectly called "overly mature forests," comes the powerful, prolonged call *kong kong kong...* that is usually heard around midnight and is the inspiration of the onomatopoeic *Mapuche* name *kong-kong.*

The Rufous-Legged Owl is identified by its varied nocturnal calls and by its brownish face with large concentric rings around its eyes. From within the dense foliage it usually emits a short call *kuhúrj* or a short, but strong, echoing shout *có-có-có*. The first sound led to its onomatopoeic Yahgan name *kuhúrj*. The second gives origin to its *Huilliche* name *coa* or *có*, which is the name used on Chiloé Island.

The *Huilliche*, who live on Chiloé Island, assign distinct meanings to this owl's calls, depending on the form. If the *coa* only shouts once, it calls for good harvests and fortune, but if it calls many times, it announces bad luck. The "remedy" to change this message of bad luck consists

of hanging salt over the wood stove.

For the *Lafkenche* the *kong kong* is a beloved bird. When a thief, or *weñefe*, comes near with intentions of stealing a domestic animal, the *kong kong* forewarns with calls that imitate a cackle, moo, neigh or bleat. As such, this owl serves as an "alarm" for the *Lafkenche* communities.

Mapuche story. For the Lafkenche, the kong kong is a nocturnal bird, very beloved for its ability to imitate other birds and animals, like chickens, cattle, horses or sheep. When some robber or weñefe comes close with the intentions to steal some of these domestic animals, the kong kong shouts. If it is going to steal a cow, it shouts ....*muuu.* If it is going to steal a horse, it shouts ....*ihhhiiiii.* If it is going to steal a sheep, it shouts ....*meheeeee*

Mapudungun version. *Tachi pu LAFKENCHE, rume ayipoyekefi tachi KONGKONG punüñüm feyta inarume pikantufilu puwera üñüm ka fentren kulliñ, femngechi achawall rume, waka rume, kawellu rume ka ufida rume. Feyti mülekelu WEÑEFE fey witramealu fill wera kulliñ ñomngelu. Fey pipingey wakeñnün KONGKONG.*
Weñengeyal WAKA..... m u u u u – m u u u u.
Weñengeyal kawell i i i i . ji- ji – ji –ji –ji
Weñengeyal ufida me – e – e – e – e

Yahgan: **Yahutéla,** Kuhúrux, Yohutela
Mapudungun: **Tukuu,** Nuco, Tuco, Ñacurutú, Raiquen
Spanish: **Tucúquere**, Búho, Ñacurutú
English: **Austral Great Horned Owl**
Scientific: *Bubo magellanicus*

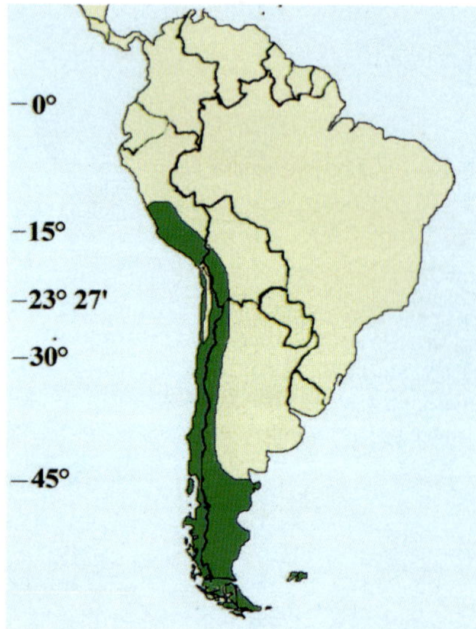

Distinguished by the horn-like tufts on its head that give it its English name, the Austral Great Horned Owl is the largest of the austral owls. Its general coloration is brownish-grey with flecks of yellow, and black horizontal bands on the underside. In the austral landscapes, it can be seen perched on branches of tall trees or flying in the search of rodents, rabbits and occasionally a bird or insect.

It emits sonorous, merry whistles when its disputes its territory or mate. Contrasting this whistle, its characteristic call from the perch is the deep voiced ululation from which it gets its Mapudungun onomatopoeic name *tukuu*. When the voice of the Austral Great Horned Owl, *tukuuhuhu, tukuuhuh,....,* is heard in the night, it seems to invoke a dense fog or *trukur* that makes the wayfarer become lost.

The Austral Great Horned Owl lives in diverse forest types and open environments, and even reaches Cape Horn, where it receives the Yahgan name *yahutéla.* According to an ancient story, they were hunters who turned themselves into owls.

In ancestral times, when birds were still people, a Yahgan child lost his father and went to live with his mother and uncle. The uncle and the other men of the camp never gave good food to this boy, and he became very skinny. One day, when she realized her son was so tired and hungry, his mother made him a pair of *kíli*, or leather sandals, and told him to go off by himself and hunt *guanacos* or *amára* (*Lama guanicoe*). After crossing a great mountain, he killed many *amára* and brought a large one back to the camp site. His mother was very happy and began to eat huge chunks of meat. Except for her, no one believed that the skinny little boy was capable of killing and transporting the *amára* to his *akar*, or hut. The next day the incredulous men accompanied the boy to the place where he left the other dead *amára*. All were surprised to see the hunted animals, and each *yamana* (man) shouldered one to carry back to the camp. Crossing the mountains the men became tired, however, because with each step the *amára* they carried pressed heavier upon their backs. Only the skinny little boy walked rapidly with his load. He was the first to return to the *akar* where his mother waited, while the tired, egotistical men advanced very slowly. Exhausted, they were just arriving around midnight, without their loads, when they turned into owls. From that time to up to our own, the *yahutéla* come around the *akars* of the Yahgans, calling in the night *tukuu, tukuu,...* only to then return to the forests, or *ashuna*.

Yahgan: **Sírra,** Síta, Lufkié
Mapudungun: **Chiwüd,** Coa, Coo, Kouw, Yarquen, Chralchral
Spanish: **Lechuza blanca, Lechuza de campanario**
English: **Barn Owl**
Scientific: *Tyto alba*

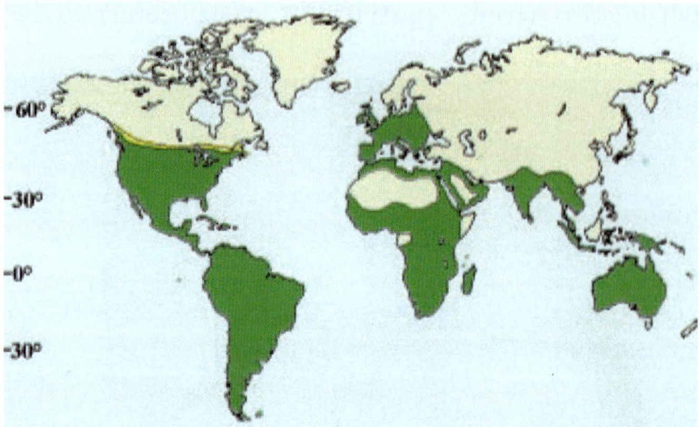

As in the case of the other owls, the *Mapudungun* name given to the Barn Owl is onomatopoeic. *Chiwüd* imitates the characteristic squeak or hiss of the Barn Owl.

For the *Lafkenche* people, this owl is an "extravagant bird" because it bandies about, giving bad directions with its squeak to travelers that try to make their way through dense fog. In the *Lafkenche* territory, it is told that when travelers come upon a *chiwüd* in the road, they are sure to become lost. Only by putting their clothes on backwards, including their right shoe on their left foot and their left shoe on their right foot, can those deceived by the *chiwüd* expect to return to their proper path.

For the Huilliche, the call of the *chiwüd* might bring bad weather, and to counteract this, people throw ashes into the air.

In the old *Pikunche* territory it is said: *chiwüd weda femkey anüpayüm* (when the Barn Owl perches on a house it brings bad omen). On the other hand, when one sees the owl hunting at dusk, it is possible to note that it feeds mainly on mice, and consequently the barn owl is beneficial to farmers. It has an excellent sense of sound, enabling it to hunt mice, even under snow.

The Barn Owl lives and nests in a variety of habitats, such as evergreen and deciduous forests, fields, rural areas and even cities. For its nests it uses tree cavities or the attic of some shed, house, church or other building. They are largely sedentary. If the pair feel safe and have food, then they do not move sites during their whole life. In the country, they are known as "monkey face" or "cat face," due to the fact that their face possesses a type of white disk in the form of a heart that contrasts with the yellowish-brown color of the rest of the head. Its flight is absolutely silent. However, upon flying it frequently emits a raspy, strident call from which it gets its *Mapudungun* name.

The Barn Owl is well known in almost all the world, and in the Americas it lives from Canada to Navarino Island. For the Yahgans the Barn Owl or *sirra* is also associated with water. It is the wise grandmother of the hummingbird *omora*, and participated in the creation of the streams and rivers of Navarino Island. Today, still the Yahgans relate the sound that the *sirra* makes when it flies at night with the sound of the streams or rivulets of water. In this way, the grandmother Úrsula Calderón tells that:

Yahgan story. "What bird is this," I asked my grandfather. "It is the Barn Owl called *sirra*; there are many of them at Santa Rosa and Mejillones. You hear them at night *shhh, shhh, shhh.*" Perhaps they sleep during the day. I saw a *sirra* when I married; it went by Mount Robalo, and I saw it. It was in a tree. So I started throwing stones. It flew away and came back again. My grandfather said that this bird, which used to sing by night, went in search for water, which didn't exist before. So this bird flew at night scattering *shhh, shhh, shhh* at all sides. Consequently, there was water in all the streams that came down the mountain. The *sirra* gave us water; water we have until now and forever.

Yahgan: **Lafkéa,** Lefkóiya, Lö(f)kw̄ia
Mapudungun**: Kill kill,** Chuco, Conchon, Chucho, Chuchu
Spanish: **Chuncho**, Caburé grande
English: **Austral Pygmy Owl**
Scientific: *Glaucidium nanum*

The Mapudungun name for the Austral Pygmy Owl—***kill kill***—repeats the sound of the owl's territorial cry: *kill, kill, kill....* When Lafkenche hear this cry at nightfall, they understand it as the announcement of the departure of a man or a woman from their home. More particularly, it expresses the fact that there are problems between the couple. If the woman is abandoned by her man, she becomes a *kazdtañpe*: a woman who will always be abandoned no matter how many times she marries. If the man is abandoned by his woman, he becomes a *lefmaufe*: a man who will never find a permanent wife.

The squawks of the Pygmy Owl at dusk and night are often raspy and strident and strike fear into the hearts of people because they resemble those of the *tue-tue*, a feared forest spirit. These nocturnal calls also inspire the Mapudungun names *chuchu* and *chonchon*. The *chuchu* inhabits all of southern Chile, and gives origin to names of places, such as Chuchunco, in the Valley of the Cachapoal river, which means water (*ko*) of the *chuchu*. The word *chonchon* describes the *winged head* into which witches transform themselves. Not surprisingly, then, the *chonchon's* night-time call is believed to announce the presence of a witch: *"kalku akuay" pikey choñchoñ*. To drive it out, they say prayers, toss salt over the kitchen fire or burn branches of the Winter's Bark Tree or *foye* (*Drimys winteri*).

The Pygmy Owl has other distinctive cries; among these is a repeated short whistle in the form of "u-u-u-u-u...". When an ornithologist imitates this owl's call in the forest, she can attract a variety of birds, which will perch nervously nearby—thus permitting the ornithologist to observe them closely.

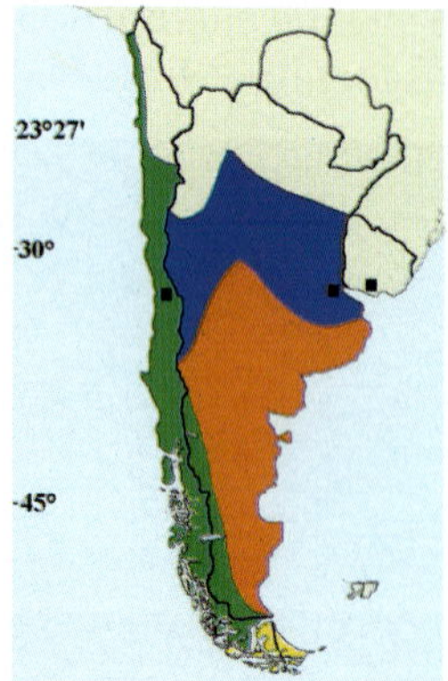

It nests in hollow trunks and while it is small and brown-chestnut in coloration, its huge eyes with yellow irises jump out at the observer day and night. It hunts at any hour; despite its small size, it is a fierce devourer of rodents, other birds and insects. For this reason, the smallest of the owls in the austral forests constitutes an important ally for country folk to control pest outbreaks.

Yahgan: **Kulálapij,** Chouhchoul
Mapudungun: **Pichi kokoriñ**
Spanish: **Peuquito,** Esparvero variado, Cazapollo
English: **Bicolored Hawk**
Scientific: *Accipiter bicolor*

The Bicolored Hawk is a diurnal raptor that inhabits the interior and margins of very diverse forests of the Neotropics, ranging from Central America to the extreme south in Cape Horn. To observe a Bicolored Hawk flying among the forest foliage is an amazing experience; the elegance and skill of its silent flight is exquisite. It is about the same size as a Chimango Caracara, but the Bicolored Hawk is distinguished by its dark grey color in the upper face, its blazing yellow irises, and its svelte body with its short, rounded wings and long white-tipped tail striped with six dark bars.

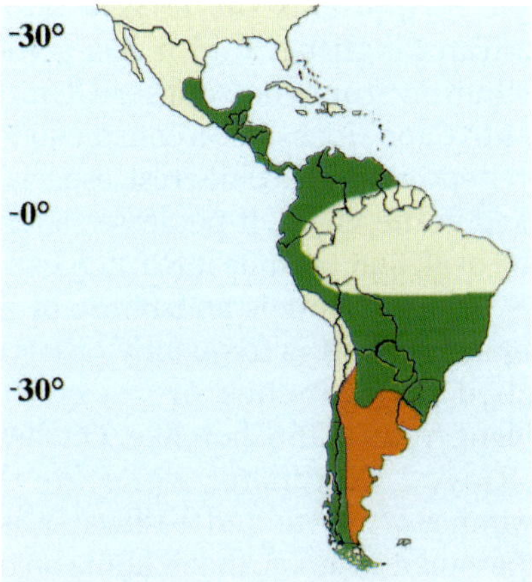

In Chile and Argentina the subspecies *chilensis* is distributed along the length of all the *Nothofagus* forests, where it hunts birds of small (e.g., Thorn-Tailed Rayaditos and House Wrens), medium (e.g., Eared Doves and Austral Thrushes), and large sizes (e.g., freshwater ducks). It also eats small rodents and insects.

Pairs nest in the canopy of tall trees, where they construct a platform about 60 cm wide over high, forked branches radiating from the trunk. They use the same nest for several years.

The Bicolored Hawk is considered a rare species, due to its low population densities over much of its range. However, it is relatively frequent in the sub-Antarctic forests of the Magellanic Archipelago. Its Mapuche name *pichikokoriñ* refers to the fact that it is a small (*pichi*) hawk (*kokoriñ*). Its Yahgan name, *kulálapij*, expresses its aggressive character, identifying it as a bad or angry (*kulála*) bird (*pij*).

Yahgan: **Chéketej,** Šégetex
Mapudungun: **Challwafe üñüm,** Queto, Quete, Kedküchan
Spanish: **Martín pescador**
English: **Ringed Kingfisher**
Scientific: *Ceryle torquata*

Ceryle torquata is the largest South American kingfisher and it has a wide distribution from the southern United States to Cape Horn. It is a conspicuously colored species with a blue crest, especially marked in the male. It possesses a long, strong beak that permits it to catch fish in rivers, lakes, channels and fjords of the extreme south. It is frequently observed perched on branches or rocks that overhang rivers or the shoreline. On these, the Ringed Kingfisher waits for the appearance of marine and freshwater fish, crustaceans and larvae that it hunts on the surface of the water or by plunging itself into it. When it notices danger, it sweeps back and forth in the air, emitting its strong, repetitive calls kekereke- kekereke- kekereke.

It nests in holes in the sides of riverbanks and hills. When she was young, Yahgan grandmother Cristina Calderón enjoyed watching the Ringed Kingfisher, or *chéketej*, coming in and out from their nests on cliffs along the Murray Channel on Hoste Island. Some old Yahgan stories related *chéketej* with another bird species that uses forests and waters, the Night Heron (*Nyctiocorax nyctiocorax*) or *huajatanu.* Both the Ringed Kingfisher and the Night Heron sleep and nest in trees, but fish in rivers or along shorelines.

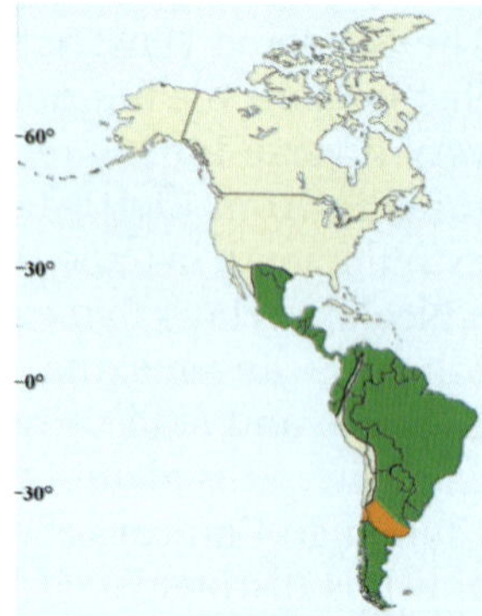

In ancestral times when *chéketej* and *huajatanu* were still humans, they were lovers. For a long time, whenever *huajatanu* left to fish in the canoe, *chéketej* helped him to catch great quantities of fish, until one day the *huajatanu's* husband surprised the lovers and impaled them with his *sírsa* or sea urchin harpoon; it was then that both transformed themselves and flew away as birds.

Grandmother Cristina was enchanted by the colors of the male and the female when they flew as a pair into the waters to

capture fish. This "enchantment" is also captured by the *Mapuche* view of the Ringed Kingfisher or *challwafeüñüm*, which is a "half tree - half water" fisher (*challwafe*) bird (*üñüm*). The *challwafeüñüm* attracts the fishes with its striking, iridescent colors that it displays in its flights skimming over the water, plunging itself into the water from sunrise to sunset. Through the hypnosis that *challwafeüñüm* exercises over the fishes, an ecological flow from the beings of the water to the beings of the air is generated. The bird's beauty, itself, forges the links in this food chain. Here, the insight of the traditional *Mapuche* ecological knowledge has essential similarities with contemporary ecological concepts of food webs.

Mapuche story. The kingfisher or *challwafeüñüm* attracts fish with its flashy colors, which it displays in its flights which skim the water. This hypnosis, that the kingfisher exercises over the fish to consume them, generates the ecological flow between the beings of the water and the beings of the air by means of the food chain.

Mapudungun version. *Fachi CHALLWAFEÜÑÜM feytañi allangechi fill wiriñ pichuñ wüñowitra mütrümkefi pu CHALLWA mülelu leufü meu üpüm ketu amuy wenteko. Fey tachi CHALLWAFEÜÑÜM mütrüm adtukey reke feyti illun challwa duamfel nümün epuñpüle üllutuwün fill kulliñ mülekelu tuwe ka kürüf meu fey ñi ipangentuaal.*

Yahgan: **Shakóa**
Mapudungun: **Porotu,** Carcaren, Cadcadeñ, Kaikayem, Kedkedeñ
Spanish: **Becasina**, **Porotera,** Becacina**,** Becacina común
English: **Common Snipe**
Scientific: *Gallinago gallinago*

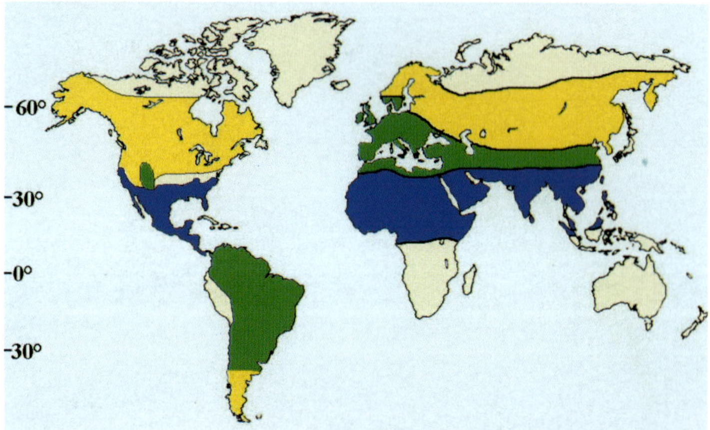

The *Mapudungun* name *porotu* is onomatopoeic with the sound that the Common Snipe male emits during the reproductive period. At dusk during spring, the male rises to the air and then falls back to the ground, making his tail feathers vibrate and generate a strange hum: "*porotu, porotu, porotu, porotu, porotu.*"

In the spring, the shrimp also make their appearance. Due to this synchrony, the strange twilight sound of the *porotu* announces harvest time for the *madeu* or shrimp, a favorite food of the *Lafkenche*. Because of its talent for extracting the *madeu* with its long beak from the caves they make in the mud, the *porotu* is also known as the shrimper-bird in the *Lafkenche* region on the coast of Temuco.

The Common Snipe is characterized by a long bill, which they use to eat invertebrates buried in the mud. Snipes live in a variety of aquatic and riparian environments, especially the inundated wetlands and rush thickets adjacent to the forests of southern Chile and Argentina.

The South American Common Snipe is considered by some scientists to be a subspecies (*Gallinago gallinago paraguaiae*) and by others to be a completely separate species, *Gallinago paraguaiae.*

Darker and more robust than the Common Snipe, the Large Snipe, *Gallinago stricklandii* (35 cm), also lives in southern South America. It is a resident of the territory south of Tierra del Fuego, where it receives the Yahgan name *shakóa*. The Large Snipe is inconspicuous because its plumage mimics the colors and patterns of the grasses and rushes, and when one approaches them they remain immobile. However, their presence is noticed at dusk during the spring, when the male elevates into the air, making his loud courtship humming sound.

Among the *Yahgan* it is not permitted to imitate the sound of the snipe or *shakóa*. She or he who imitates the sound of the *shakóa* will awake with her or his toes cut by blades of *ushkulampi* or rush (*Marsippospermum grandiflorum*) that grow in the bogs. In fact, the grandmothers Úrsula and Cristina Calderón remember that when they were little girls one day they awoke with the big toe of each foot cut because of imitating the *shakóa*, even when they had covered their feet well. They tell that:

Yahgan story. My father and my grandparents and my mother said that one should not imitate the snipe or *shakóa* when it is flying during the night. "If you imitate the bird, they will hurt you. They will make a cut in your big toe with grass or *kuruk*," they said.

As we were naughty and we wanted to test if the things were certain or not we decided to try it. We were three little girls. So we wrapped our feet with a very thick cloth and with a sock over it, and began to imitate the bird at night. The next day we woke up with a lot of pain in our feet, because the story of the snipe was true. We were certainly naughty girls, but after that we never did it again because we knew that it was certain what the grandparents told us.

Yahgan: **Waior,** Heuhs, Wéásh
Mapudungun: **Pideñ,** Piden, Cotuta
Spanish: **Pidén,** Cotuta, Gallineta común
English: **Plumbeous Rail**
Scientific: *Pardirallus sanguinolentus*

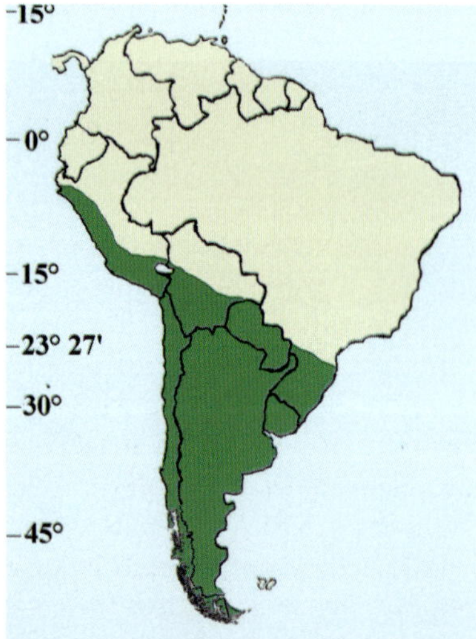

The strident call of the Plumbeous Rail characterizes the swampy landscapes in the south of Chile and Argentina, and gives rise to the *Mapudungun* name *pideñ*, onomatopoeic with its loud call *pidreeen, pidreeen, pidreeen*. This noisy bird is found from the middle of South America to its southern tip, inhabiting bogs, wetlands, river banks, streams and lakes from a great variety of ecoregions. Among the rushes they move with skill, placing one leg before the other in such a way that tracks are left in a single line.

Its legs and eyes are an intense red, while its body is somberly-muted with a

blackish-blue chest and an olive-brown back. It has a long, multicolored beak (blood red at the base, bluish in the middle and a greenish tint at the tip) which it uses to feed upon a large variety of invertebrates and vegetable material. In order to be heard in the dense vegetation, the Plumbeous Rails emit a strong, prolonged series of cries, especially at dusk when they listen to the answers of different members of the population.

For the *Mapuche* people, the *pideñ* is a bird that forecasts the weather with its calls at the end of the day. On Chiloé Island, the *Huilliche* and farmer communities predict that if the rails sing in chorus at sunset, the following day will have good weather. The *Lafkenche* say that if the *pideñ* launches its calls towards the north, there will be a torrential downpour; if these go towards the south there will be good weather. As the Lafkenche poet Lorenzo Aillapan recites:

Mapuche story. The strident call of the *pideñ, cotuta,* or *gallineta* characterizes the swampy landscapes in the south of Chile and Argentina. As the bilingual poem *Pideñkawun* says:

The resounding cry of the Plumbeous Rail
Gathered together they sing, and from the matorral they depart,
announcing bad weather and rain.
In the dense river-cane they hide,

wading birds with red legs,
beaks similar to the Oystercatcher.
¡Piiiir with piiiir with fiiiir with fiiiir with piiiir with piiiir with fiiiiir with fiiiiir with!

Mapudungun version. *Tachi waken ülkantun pidén, cotuta achawall pingey kimfaluwi rumel focholen adlelfün willipüle Chile/mapu ka Argentina/ mapu. Feypiley tachi epu rume dungun koyautun. Pideñkawün meu:*

Pideñ tañi fillke kümüllün wakeñnün
Fentren üñüm re kefafan tripapay walfe mew
Rüf küftukun mawünay pileyngün
Cheu ñi mülen trongen külantu ellkaupuyngün
Epu kelü namun tuley rofi küpal üñüm
Taiñ komilwe wün pilpilen üñüm femngey.
¡Piiiir with piiiir with fiiiir with fiiiir with piiiir with piiiir with fiiiiir with fiiiiir with!

Yahgan: **Tulára-Táchij**
Mapudungun: **Chiuchiu**, Chiuchihuen
Spanish: **Churrete acanelado**, Remolinera común
English: **Bar Winged Cinclodes,** Dusty Cinclodes
Scientific: *Cinclodes fuscus*

Besides the Bar Winged Cinclodes, there are several other species of Cinclodes that have similar behavior and calls. Consequently, they also receive the name of "remolineras" or tail-bobber, and the *Mapudungun* onomatopoeic name *chiuchiu*. One of the species distinguishes itself from the others by inhabiting the shoreline of rivers all the way to the tops of the mountains. For this reason, its Yahgan name is *tulára-táchij*, which refers to the Cinclodes (*táchij*) that principally inhabits the mountains (*tulára*). The coloration of this Cinclodes is also brownish or cinnamon, which is more mimetic of the mountain environment.

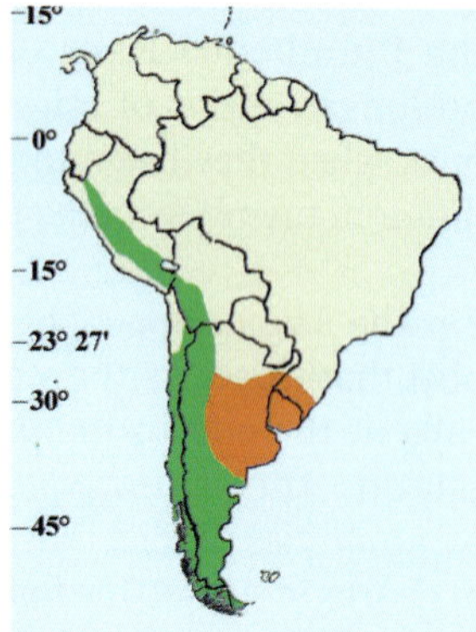

Cinclodes might surprise us: the Blackish Cinclodes (*C. antarcticus*). This is the only species of cinclodes without a wing bar. Its pre-ferred habitats are the grass-lands of the exposed islands and coastlines it shares with colonies of sea birds and sea lions.

In keeping with the lifestyles of the other Cinclodes species in the extreme austral, it is not rare to find the Bar Winged Cinclodes marau-ding in the pools and wet areas in the interior of the evergreen forests. But, in the Mage-llanic Evergreen Beech forests and other coastal eco-systems of the Yahgan territory, an endemic species of

Yahgan: **Táchij**
Mapudungun: **Chiuchiu**, Churrete, Churreta, Churete, Chiuchihuen, Thureu, Huechuthureu
Spanish: **Churrete**, **Remolinera araucana,** Remolinera común, Meneacola, Piloto
English: **Dark-Bellied Cinclodes**
Scientific: *Cinclodes patagonicus*

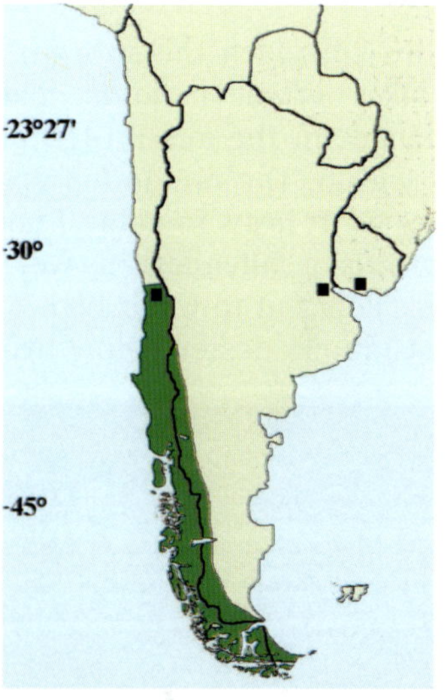

All the Cinclodes species present similar behaviors and vocalizations. Hence, all of them receive the same common Spanish names, and the onomatopeyic *Mapudungun* name *chiuchiu*, which derives from the bird's short vocalizations *chrrr-chrrr*. They fly from rock to rock, moving without stopping. They only come to rest for a brief instant to look in every direction while they bob their tail and emit their short call *chrrr-chrrr* or *chiuchiu*. From this behavior comes the Argentine names *remolinera* or *meneacola* (tail-bobber). In contrast, its Chilean name, churrete, is derived from the fact that these birds have a very active digestion, and leave behind large stains

of abundant feces over the rocks and the shore. In Chilean slang "to be *churrete*" means to have diarrhea. Finally, the name *pilot* given to the species of Cinclodes by fishermen and sailors from the channel and fjord region of the south comes from the white markings around the eyes that resemble navigator's goggles.

When we approach the shorelines of rivers, lakes, as well as fjords, channels and beaches in the austral temperate forest region, the cinclodes species surprises us. It nests in natural hollows among rocks or fallen logs, as well as in buildings adjacent to water bodies, and is an assiduous nester below bridges. Grayish-brown in color, it mimics the rocks and stones where it hops about in search of all types of crustaceans, mollusks and invertebrates. The only showy parts of this bird are its eyebrow that extends to the nape of the neck, its pale throat and its song.

Along coastal shorelines and rivers its short vocalization, *prrr, prrr, prrr*, is characteristic as its skips from rock to rock. During the reproductive season it is common to hear the cinclodes sing its melodious, sharp, strong and prolonged trilled song, while it beats its wings perched on an overhanging branch or rock.

Yahgan: **Lásij,** Lášix, Lášixkipa
Mapudungun: **Pillmaykeñ,** Wüshükon, Pillmayken
Spanish: **Golondrina chilena**, Golondrina patagónica
English: **Chilean Swalow**
Scientific: *Tachycineta leucopyga*

The swallows of the forests of southern Chile and Argentina include several migratory species that arrive in spring and leave in fall. Among them, the Chilean Swallow and the Blue-and-White Swallow are the most common. Both species receive the Mapudungun name *pillmaykeñ*, which means spirit (*püllü*) that travels (*may;* the suffix *keñ* makes the name a noun), and their arrival announces the spring. The Yahgan language also assigns the same name, *lasij*, to both the Chilean Swallow and the Blue-and-White Swallow.

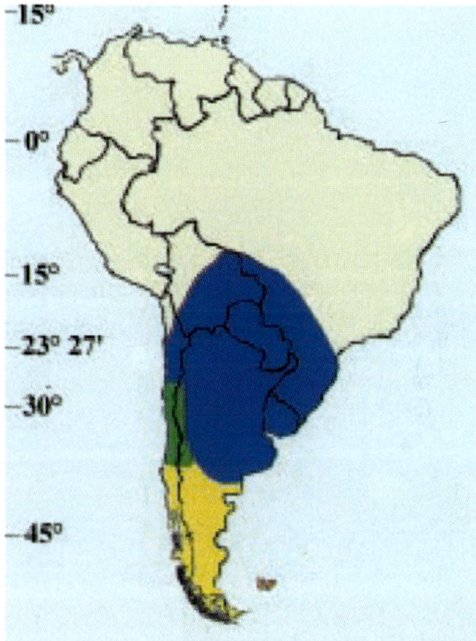

Sometimes the Chilean Swallow was called *lasijkipa* in Yahgan because it was considered one of the wives (*kipa* = women) of the Wren or *chílij* (*Troglodytes aedon*), another small bird with migratory habits. This pairing was probably inspired by the observation that, in the extreme south, *lasijkipa* and *chílij* both nest in cavities of banks or hollows in standing dead or live trees. These species take advantage of naturally occurring cavities and those made and abandoned by woodpeckers.

Swallows play an important role in the control of pests in the forests and agricultural fields, given that they fly tirelessly hunting for insects. That is why they are frequently seen flying artfully and doing pirouettes; their aerial dance allows them to consume millions of insects. The *Huilliche* communities of Chanquín and Huentemó south of Chiloé National Park carefully observe the type of flight of *pillmaykeñ* in order to forecast the weather: if the swallows fly high they announce good weather; if their flight is low to the ground, the weather will be bad.

The Chilean Swallow is identified by the bluish-black color on its back which is interrupted by a white stripe before the tail. Its chest is also white.

In the *Mapuche* territory, these swallows provide the source of the name of the beautiful Pilmaiquén River

(*Pillmaykeñlewfü*) with its famous waterfall. The Chilean Swallow nests in the roofs of houses in the cities and countryside of Chile. Its arrival announces springtime and romances. That is why many girls are named *Pillmaykeñ*, as told in the following story recorded with the poet Lorenzo Aillapan.

Mapuche story. The Mapuche name for the swallow is *pillmaykeñ*, which means spirit that travels, flying about the earth. Its arrival announces the spring, that is, the beginning of the period of courtship. Hence, the swallow excites the young people who frequently are given the name *pillmaykeñ*. During the 1990's, the tradition of giving sons and daughters bird names has revived in the Mapuche communities. Today, the name *pillymaykeñ* is as popular, or more popular, among little girls as the Christian name María.

Mapudungun version. *Tachi üytukun pillmaykeñ tripalu püllü kañpüle ka fey akutulu weflu pewü, müleyüm dakelüwün duamtu, fey pu wekeche ülcha deuma pillmaykeñ üytuy. Fey pu lofche, rupalen tripantü 1990, pu kurewen che femngechi pu fill üñüm üy tukutuy ñi pu fotüm ka pu ñawe. Femngechi, lofmeu kiñeke doy allmantuy tukual pillmaykeñ üy chumngechi mülefuyta wera fentren ka feyentun pu malen "María" pingekefulu.*

Yahgan: **Lásij**, Lăsix, Lăsixkipa, Ƚasix
Mapudungun: **Pillmaykeñ,** Wüshükon, Pillmayken
Spanish: **Golondrina de dorso negro**, Golondrina barranquera
English: **Blue-and-White Swalow**
Scientific: *Pygochelidon cyanoleuca*

The Blue-and-White Swallow is another common swallow species in the forests of southern Chile and Argentina. It is more selective with respect to its preference for aquatic habitats than the Chilean Swallow and has no white band on its back. The Blue-and-White Swallow abounds near bodies of water where mosquitoes and other insects thrive. As they fly about in curling masses, their polyphony of voices suggests, as Lorenzo Aillapan indicates, a group engaged in "birdly conversation."

The Blue-and-White Swallow is migratory and covers a wide range of territory from Mexico to Isla Navarino, Chile. In the fall, it migrates north and returns each spring to the austral forests where it is almost always seen chasing insects over the rivers, lakes and even fjords.

The Chilean Swallow's reputation for being a good huntermight be related to an ancient Yahgan narrative. In ancestral times, when birds were still humans, the swallow or *lasij* was a small hero who jumped into the mouth of a giant dolphin, which swallowed it. Once it was inside the enormous dolphin, the little *lasij* used a sharp knife to cut the internal organs of the marine mammal. *Lasij* needed a great deal of time to kill the dolphin. Only when *lasij* began to hear petrels and albatrosses flying outside did the hurt dolphin swim exhausted towards the beach. There he died and provided abundant meat for all the Yahgans. Thanks to *lasij*, the Yahgans had food that permitted them to hold the celebration of the *kina*, an important initiation rite of young boys, which had not been possible in a long time.

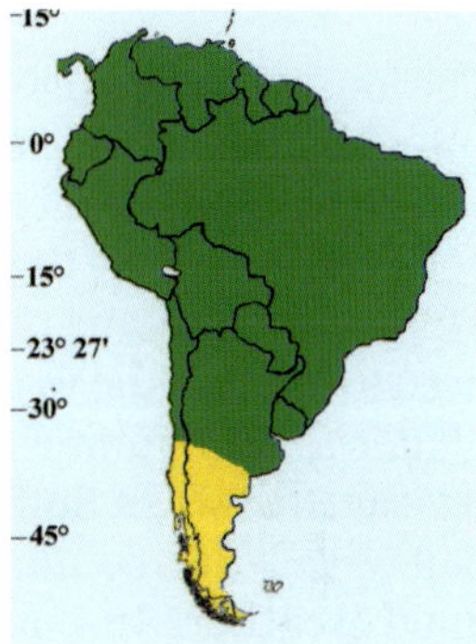

Yahgan: **Lejuwa,** Léxuwa
Mapudungun: **Raki,** Raqui, Raquin, Rakiñ
Spanish: **Bandurria,** Bandurria austral, Bandurria común, Bandurria baya
English: **Buff-Necked Ibis,** Cinnamon-Necked Ibis
Scientific: *Theristicus melanopis*

The loud metallic calls of the Buff-Necked Ibises distinguish the temperate forests of southern South America. Thest strong vocalizations lead to its onomatopoeic *Mapudungun* name *raki*. These calls are used to forecast the weather by *Mapuche-Huilliche* communities. On Chiloé Island it is said that when Buff-Necked Ibises call half way up the mountain, they announce good weather. In contrast, when they shout with their wings open, perched on the trees or crow in chorus, they announce bad weather. When they cry at the moment of taking flight, then there will be a south wind and the weather will get better.

Moving in flocks, the ibises clean the fields of insects that could be harmful to the plantings. In the fields and other open habitats it is common to observe numerous ibises scratching with their long beaks in search for earthworms, larvae, snails and other invertebrates. To sleep and nest, they use the canopy or branches of tall trees, where they come together in colonies.

At the end of fall, the populations migrate to the north along the length of the Andes, returning to the south to announce spring. The Yahgan culture also associates the migratory movements of the Buff-Necked Ibis or *lejuwa* with marked seasonal or climatic changes, as is appreciated in the following history narrated by the Yahgan grandmother Cristina Calderón.

Yahgan story. In the times of the ancestors, one day when spring arrived, a *yamana** leaned out of his *akar* or hut and saw a Buff-Necked Ibis or *lejuwa* flying in the sky. The *yamana* was so happy that he yelled to the other members of the community, "An ibis is flying over our *akar*. Look!" Immediately the others came out of their huts, shouting loudly, "The spring has come, the ibises are already in their return flight." They jumped for joy, and they jumped hard.

Lejuwa is, however, a very delicate and sensitive woman and likes to be treated with special deference. Upon hearing the noise of the shouts, she became furious, and profoundly offended, causing a strong snow storm. It snowed incessantly; it was freezing cold; all the land was covered with ice; and all the waters froze. Many people died because they could not navigate in their canoes to search for food. Neither could they leave from their *akars* to look for firewood because all was covered with snow. More and more people continued to die. Only after a long time did it stop snowing, and the sun begin to shine.

The sun warmed so much that begin to melt the ice and snow that had completely covered the land, and consequently floods began to flow towards the channels and the oceans. The ice that covered the wide and narrow channels broke apart and melted, and the Yahgans could finally go to the shores and navigate in their canoes to collect food. On the big mountainsides and deep valleys, however, the ice was so thick that not even this sun could melt it. Even today, glaciers can be observed coming down to the sea, that remind us of the severe frost and snowstorm that *lejuwa*, who is a very delicate and sensitive woman, provoked in those times. Since then, the Yahgans treat the Buff-Necked Ibis or *lejuwa* with much respect, and when they come near the houses or *akars*, the people keep quiet, particularly the little children who are not allowed to imitate them or to cry.

*Yamana *is an ambiguous term because this is the name of the ethnic group and refers to a Yahgan man. In this case it refers to the second meaning.*

Yahgan: **Pílit**
Mapudungun: **Tregül,** Queltehue, Queltreu, Treile
Spanish: **Queltehue**, **Treile, Tero,** Tero-tero, Fraile
English: **Southern Lapwing**
Scientific: *Vanellus chilensis*

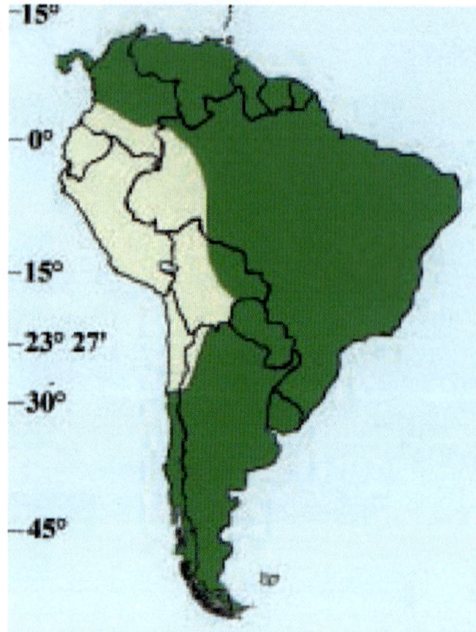

With its strong cries of *tregül, tregül, tregül* that provide its onomatopoeic *Mapudungun* name *tregül*, the Southern Lapwing is a bird that characterizes diverse open habitats. This regular-sized bird is common throughout South America, except Amazonia. In southern Chile it is common to observe the Southern Lapwing in the grasslands and wet fields adjacent to forests, living together with flocks of Buff-Necked ibises and other birds, whil feeding from the soil on every type of invertebrate including larvae and adult insects.

The Southern Lapwing carries on its head a long and thin black crest. Its body stands out by being multi-colored: *i*) it is black in the front, the lower neck, chest and external tail-coverst; *ii*) light grey on the rest of its head and neck; *iii*) greenish-brown with a metallic shine on the upper parts of the wings; *iv*) white on the flanks and sub-terminal feathers; *v*) pink on the legs and beak; and *vi*) the iris of the eyes is an intense red. The Southern Lapwing displays all its coloration in its courtship flights and dances while it emits varied calls. This courtship display of the *tregül* inspires the *Mapuche* rite of the *ngillatun*.

The *tregül* are welcomed in the *Mapuche* and farmer communities of southern Chile because they are meticulous "cleaners." With their pecking and their dung, they *tregül* also contributes to the fertility of the soil. In addition, their flight forecasts the weather: when the *tregül* flies low it is certain to rain.

The *tregül* is normally calm, but it becomes very aggressive when defending its nest or territory with its strong barbed wings. *Huilliche* people call it the "vigilant bird" because it announces the presence of any person or animal passing through the area. The *tregül* screams whenever it sees someone, even in the middle of the night, and makes one *Lafkenche* and *Huilliche* people wonder if it ever sleeps. As the Mapuche Bird-Man says:

Mapuche story. The Southern Lapwing is a sentinel, who upon observing raptors and other hunting animals, alerts the other bird species with its call *tregül, tregül, tregül....* Well armed with spurs on its wings, the Lapwing confronts the foreigners in its territory, but it also strongly expresses romance, its love of life, with a variation of its call: *triliu, triliu, triliu, triliu, trültriu, trültriu, trültriu, trültriu, trültriu.* With this call the male courts the female, dancing and singing. This courtship is represented in the Mapuche rite of *ngillatun*, where with open wings, like open arms, it dances counter-clockwise, calling out *trilililiiliuuuuuu* towards the north, *trilililiiliuuuuuu* towards the west, *trilililiiliuuuuuu* towards the south, *trilililiiliuuuuuu* towards the east. With this romantic song and its lover's dance the Southern Lapwing invokes the Universal Fertility Spirit.

Mapudungun version. *Tachi tregül (tero) lloftukellufe üñüm ka feyti kulliñ femngechi miyaukelu tralkatufe lelfün meu tie küpay pipingey tregül, tregül, tregül...... pu mürwen wera üñüm ellkautuy. Müna küme waykituley tañi epu müpü, tregül may kewa wemükefi kayñe akukelu ñi ad mapu lelfün meu. Welu kafeyti tregül duantulu pünowal müna küme ül nentuniekey fill wünül: triliu – triliu – triliu... trultriu, trultriu, trultriu... Femngechi feyti alka pünoy kude tregül pürüy ka ülkantuy. Katachi kawin dungu amuley feyñi chumken mapuche NGILLATUN, kefafay ka amuldunguy epu lipan nülay (müpü reke nülay) fey amuley rakiñ antü reke, wünül dunguley trilililiiliuuuu pikum püle trilililiiliuuu lafken püle, trilililiuuuu – willi püle, trilililiuuuu tripan antü püle. Omfucha/Omkude mütrüm adtuleyngün.*

Yahgan: out of bird´s range
Mapudungun: **Kono,** Konun, Turcasa, Cono
Spanish: **Torcaza**, Paloma araucana
English: **Chilean Pigeon**
Scientific: *Columba araucana*

In the forests of southern Chile and Argentina inhabits the Chilean Pigeon or *kono*, an endemic pigeon larger than the domestic one so common in the world's cities. *Kono* has a beautiful, reddish-chestnut coloration, orange eyes, and an elegant white band at the nape of the neck with a metallic green patch below. It is gregarious, and lives in flocks high in the trees where they eat fleshy fruits like the Looser (*Cryptocarya alba*), the Lingue (*Persea lingue*), the Winter´s Bark (*Drimys winteri*) or the *Olivillo* (*Aextoxicon punctatum*). They nest in the trees, constructing small platforms of small, dry sticks, where they incubate and then feed their chicks with a kind of "milk" digested from the seeds of the fruits.

Hidden in the foliage of the trees, the pigeons emit their sonorous cooing that so typifies the austral forests—the sound listened to by Spanish conquistadors and caused them to believe that it was the most abundant bird. Places such as Conumo (37°16'S; 73°14'W) in the mountains near Arauco and the town of Pucón (39°15'S; 71°58'W) on the shores of Villarrica Lake express with their names the residence of the Chilean Pigeon. Conumo means "with (*meu*) pigeons (*kono* or *konun*)", and Pucón denotes pigeons (*pu* = plural prefix; *kon* = abbreviation *kono* or *konun*). As described by the poet Lorenzo Aillapan, the Chilean Pigeon or *kono* was so abundant in past times that its flocks covered the sky with winged clouds—clouds that floated through the *ankawenu* or celestial space of the Mapuche territory.

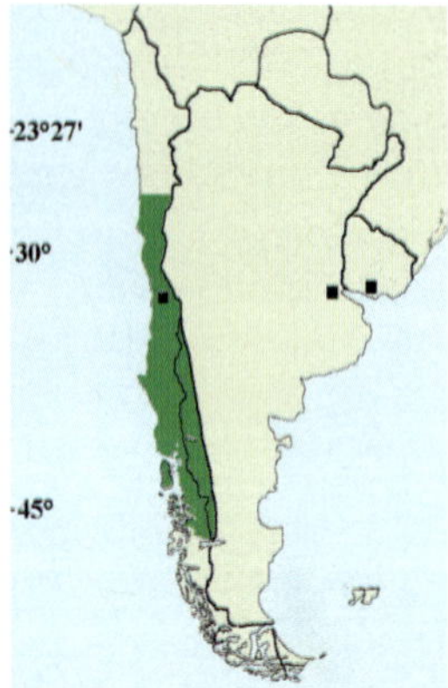

During the first decades of the 20th century , the Chilean Pigeon became the favorite prey of modern hunters in southern and central Chile. They waited to hunt it in the month of April, when the *lingue* (*Persea lingue*) bears fruit. When pigeons gorged on the *lingue* fruit, its flavor permeated their flesh, a flavor highly-prized by themodern hunters and their customers. The over-hunting that ensued to satisfy the appetite for this flavor, together with habitat degradation and the devastating effect of the Newcastle epidemic virus, carried the Chilean Pigeon to the edge of extinction. Fortunately, there is still hope: *kiñekelewey kono mawida mew*, there are still a few wild pigeons flying in the austral forests.

Today its hunting is forbidden and the populations of Chilean Pigeon have begun to recuperate. It is possible to observe again flocks of more than 30 individuals. Even more important, the existence of *kono* and its habitat are beginning to be respected again, and the diverse, human cultural viewpoints in southern Chile and Argentina are being revalued today.

Yahgan: out of bird´s range
Mapudungun: **Maykoño,** Cullpo, Culpo, Mayconu, Muikoño, Cono
Spanish: **Tórtola**, Torcaza
English: **Eared Dove,** Spotted Dove
Scientific: *Zenaida auriculata*

According to the *Mapuche* view, this small grey dove with black spots and bands on the wings is one of the half forest (mawida) and half prairie (lelfün) birds. Every sunrise, flocks of Eared Doves fly from the woodlands to the prairies, feeding places where they search for any type of seed and for ponds to drink water. Each sunset the flocks return to their trees in the woodlands, where they have their nesting and roosting site.

This daily cycle creates Eared Dove flyways that the hunters know very well to wait with their rifles for the flocks that pass flying in the morning and late afternoon of each day. Fortunately, the Eared Dove is very prolific with up to three clutches per year, and their populations remain abundant in Chile and Argentina, as well as almost all of South America (except the Amazon Basin and the Andes above 3,000 m). It even has become a frequent inhabitant of villages and urban centers. The Eared Dove is today a native pigeon in the central plazas of cities like Temuco or Chillán in the south of Chile, or in Quito or Bogotá in northern South America.

The Eared Dove, as with the domestic pigeon, dedicates much of its time to reproduction. It constructs its nests of little sticks and feathers, which often fall onto the soil. This circumstance reflected in the *Huilliche* name of Muicolpué, in the Coastal Mountain Range near Osorno, which indicates place (*ue*) of dove (*muiko*; abbreviation of *muikoño*) feathers (*lp*; abbreviation of *lepi*). As such, *maykoño*

accompany the rural and urban human communities, and from their nests and roosting sites comes their characteristic song, the typical cooing of the doves. Its call and the long periods it spends on its nest are poetically described and interpreted by the Lafkenche poet Lorenzo Aillapan.

Mapuche story. Maykoño is a feminine name, and with its pitiful call (*uh, uh, uh, uh, uh,...*), the Eared Dove has cried since humanity has existed. It is a spokesperson for the pain of existence, suffering for the frequent fall of its little eggs or chicks from the rudimentary nests, constructed with a few little branches. It cries with the pain of other birds, other living beings and of the Earth itself, hurt by human actions, such as the extensive clear-cutting of forests and the excavations of mines. It cries for natural furies, like volcanic eruptions, earthquakes and tidal waves. The dove or *maykoño* reminds us with its cooing (*uh, uh, uh, uh, uh,...*), that pain is as integral a part of life as is happiness and pleasure.

Mapudungun version. *Tachi maykoño domo üyngey ka niey ngüman ülkantu: uy-uy-uy-uy kogon-kogon kogon-kogon uy-uy-uy-uy. Trangoy-kuram-püñeñ-kogon. Tachi maykoño ngümakey deuma mülen mogen che. Ülkantun meu nentuniekey mongen duam, trangoy kuram, trananagita püñeñ rume pichin kochay niey ñi dañe, katachi kutrantulelu ka wera üñün, kulliñem mongelelu katachi ñukemapu allfülngelu fillfemün pu wedake che, katrün mawidantu ka kintugen fill duamfal pañillwe, ka fey illkun ad mapu, kütral degiñ, fütranüyün ka tripako lafken. Fey ngümay weñagütu tukulpay küme ül uy-uy-uy-uy.... kutrantunlladkün weraley mongen fey ka wefrumetuy ayiwün ka llakong duam.*

Yahgan: out of bird´s range
Mapudungun: Pütriu, Pitio, Pitihue, Pitiu, Pitigüe, Rere, Pütiw
Spanish: Pitío, Pitigüe
English: Chilean Flicker
Scientific: Colaptes pitius

The Chilean Flicker is the most common of the Chilean woodpeckers, and typifies the diverse forest and shrubland habitats with its resounding territorial call *pitrío, pitrío, pitrío ...* This call gives origin to the onomatopeic *Mapudungun* name *pütriu*.

The *pütriu* occupies a great variety of habitat types—in fields with few trees, in regenerating forests, and in old-growth forests—from sea-level to 2,000 m elevation. In the forests the flickers also devour berries and seeds, especially from the *quilas* or bamboos (Chusque sp.).

In contrast with the other austral woodpeckers, the *pütriu* feeds not only on trunks, but also in the soil where it captures a great variety of worms, larvae, eggs, nymphs, and adult insects, including ants. To nest, it excavates cavities in both old trees and earthen banks. Like the Eared Dove or *maykoño*, the *pütriu* is a half-tree and half-soil bird, and from the trees it announces danger with its alarm call. As the *Mapuche* Bird-Man, Lorenzo Aillapan says:

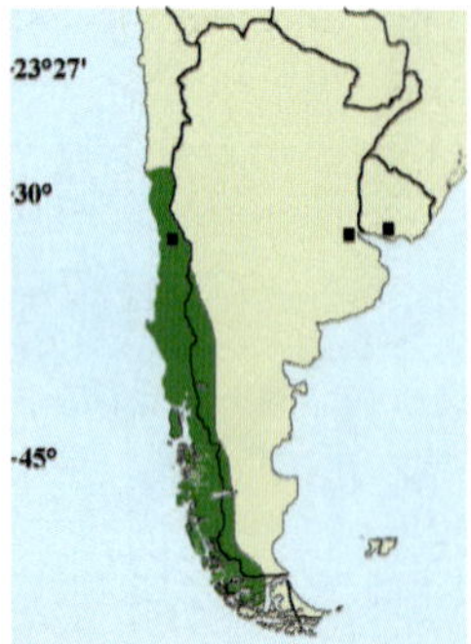

Mapuche story. The *pütriu* notifies the other birds with its shout of alarm *trü trü trü trü trü* the presence of the hunter. With its other shout *trü tü tü tü tü* it notifies that the hunter has gone and that the half tree – half ground birds can return with safety to the prairie. As such, the communication of this carpenter bird expresses the spirit of solidarity that exists between communities of birds and which beautifies Mother Nature.

Mapudungun version. *Tachi pütriu inkaniekey pu mürwen üñüm wakeñ ül men pütriu-pütriu-pütriu-pütriu-uuuu... tie küpaytralkatufe, feytañi wakeñ ül meu: pütriu-pütriu-pütriu-pütriu-pütriu-uuuu tie amutuy tralkatufe feula wiñotuay pu wera üñüm rangiñ mawidantu ka rangiñ lelfün küme pitrongtuay wera kachu lelfün meu. Feymeu tachi katafealiwen üñüm kiñewayiñ dungu mingako püllü weftuay taiñ üñüm kiñewayiñ dungu mingako püllü weftuay taiñ üñümngen ka fey meu rume adtuay taiñ ñuke mapu.*

Yahgan: out of bird´s range
Mapudungun: **Pichi rere**
Spanish: **Carpinterito**, Carpintero bataraz grande
English: **Striped Woodpecker**
Scientific: *Picoides lignarius*

The ancient *Mapuche* called the family of the woodpeckers *katanaliwenüñüm*, alluding to the birds' (*üñüm*) habit of perforating (*katan*) the trees (*aliwen*). The Striped Woodpeckers are the smallest representatives of this group of birds, from which comes its name *pichirere* (*pichi* = small; *rere* = woodpecker). Its scientific name, *Picoides lignarius*, means "similar to the woodcutter" and this little woodpecker is, indeed, recognized by its rapid and constant pecking. In the countryside, people know it as "wood chopper."

The male has a characteristic red forehead and crown. He marks his territory with a rapid succession of pecks against a dry trunk, producing the sound that ornithologist Guillermo Egli describes as "drumming." To build their nests, Striped Woodpeckers excavate small cavities in the trunks of old, living or dead standing trees.

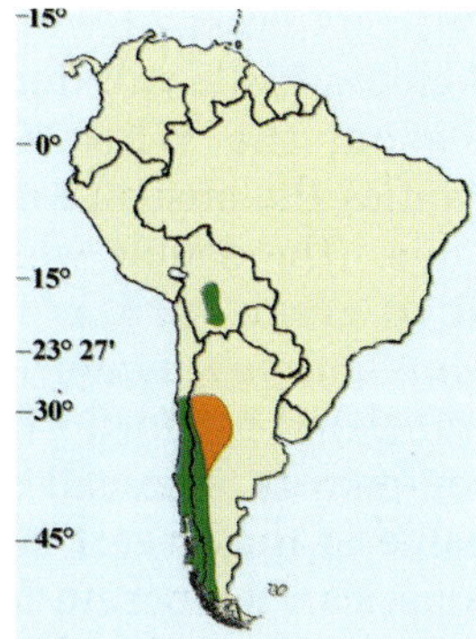

In many types of forest, the *pichirere* can be seen going laboriously up and down tree trunks, supported by their tail and extracting larval and adult insects with its strong beak. The Mapuche Bird-Man tells the story of this member of the *katanaliwenüñüm* family:

Mapuche story. Striped woodpeckers or the *pichirere*, the same as the *pütriu* and the *rere*, pass the day pecking on the trunks of trees, extracting the larvae, which likewise feed on the sap and the wood of the plants. As such, the flow of life is established between the sap of the plants, the larvae of the insects and the woodpeckers, who resound in the forests, or *mawida*, of the *Mapuche* lands.

Mapudungun version. *Tachi pichi katafealiwenüñum pichirere, ka feyti pütriu ka ti rere, rumel pünalekey pitrong-pitrong aliwen mamüll, fey nentual piru merun fid mongepiyüm fill aliwen ka añumka. Femngechi, epuñpüle mongeluwi tuwikülenmeu fill anümka ka feyti wera piru fill idike fey wepumniey tachi pu katafealiwenüñum fey negümnegümngey mawida meu cheuñimülen pumapuche.*

Yahgan: Feo, Fuiiu, Puiu, Pouyou
Mapudungun: **Wiyu,** Fio, Fiu-Fiu, Fio-fio, Vio-vio, Peutren
Spanish: **Fío-fío**, Fío-Fío silbón
English: **White-Crested Elaenia**
Scientific: *Elaenia albiceps*

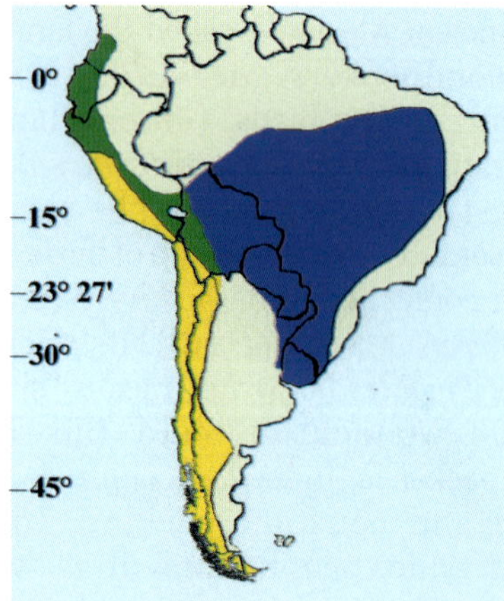

With its crest of white plumage and double-whistle call during the breeding season, the White-Crested Elaenia typifies the austral forests of southern Chile. The double whistle of this little, olive-green bird is imitated by the *Mapudungun* name *fiu-fiu*, which was translated as *fío-fío* by the Spaniards conquerors. The call also inspires the name of the largest river in the South American temperate forest region, the Bío-Bío River. The single whistle of this bird is heard in the early morning and gives rise to the other *Mapudungun* name: *wiyu*.

White-Crested Elaenia stands out among austral avifauna for its remarkable migratory behavior. It spends the winter in tropical cloud forests of Ecuador, Peru and Brazil. At the beginning of spring adults begin their flight south and go as far as Cape Horn, where they nest and feed their young, which then fly with them back to the tropics at the end of summer. The migratory flight of the White-Crested Elaenia is so precise that, even having flown 5,000 kilometers, they arrive to southern Chile and Argentina and nest in the same trees year after year. It often builds its nests and feeds on the fruits of a tree that is sacred to the *Mapuche* people, the *foye* or Winter's Bark (*Drimys winteri*).

The White-Crested Elaenia is a diet generalist. Throughout the year it is a skillful insect hunter, but during the flowering period of the *notrü* or firebush (*Embothrium coccineum*) it acts like a

hummingbird, flying from flower to flower, drinking nectar. When the plants begin to fruit in summer, it eats the berries of many trees and brushes, such as Box-Leafed Barberry and Wild Currant.

In the forests and mountains of the Mapuche territory, where berries abound, the whistles of the *wiyu* also inspire many entertaining stories that are told between women. Poet Lorenzo Aillapan tells us that:

Mapuche story. Near the coast of Temuco, the Lafkenche associate the whistle of the White-Crested Elaenia, the *wiyu*, with the whistles that young men direct towards women when they go to collect *maki* (*Aristotelia chilensis*) and other fruits in the forest. It is told that some women have recovered from sickness when they receive the love of these young men. Humorously, a beautiful grandmother from Cunco, towards the mountains from Temuco, went one morning to collect water at a well where she heard these flirty whistles *fio, fio, fio...* Amazed she said, "it seems that I am still attractive." She emptied the bucket of water and returned to the well right away, where she only found not a young lover but the roguish *wiyu* instead, whistling its characteristic call.

Mapudungun version. *Pu lafkenche, konün antü püle Temüko waria, keñangekey wiykeñ wiyu pengeyüm fütra pura domo miyaukelu domokulme wechewentru (maqui) makemeyalu fill pu domo mülekelu kam fill fün mawida. Fey piam kiñe ülchadomo llaftuy fechiwüla wentrutulu, rumel tremotuy - fey may pi: wentru mayta lawenwürke. Allangechi, kiñe kuku mülelu kunko (Cunco) waria püle, pirentu winkul Temüko püle, kiñe liwen amuy komealu kotuwe meu fey allkütuy wiykeñ wentru domokulngen. Feynga konduamlu feypi "Petu may illufalngeperkelan küme trawa kam nielu inche", witruy ñi medenko fey ka wuñotug tañi kotuwe meu, welu feynga rewedwed wiyu üñüm fey wiykeñ wiykeñ ülkantu mekerkey.*

Yahgan: **Twín,** Tuwín, Dw̄iin, Dūiin
Mapudungun: **Chidüf,** Quechan, Chirihüe, Dihue
Spanish: **Jilguero**, Cabecitanegra austral
English: **Black-Chinned Siskin**
Scientific: *Carduelis barbatus*

The Black-Chinned Siskin stands out due to the male's prolonged call and showy colors. The male's black crown and throat are denoted by the scientific name *Carduelis* (siskin) *barbatus* (bearded), and in its Argentine name "little black head." Before and after the nesting season, they form large flocks that bring music to the austral forests as far south as Cape Horn. In the Yahgan territory, stories of the siskin or *twin* relate the beauty of these little birds to their descent from young maidens:

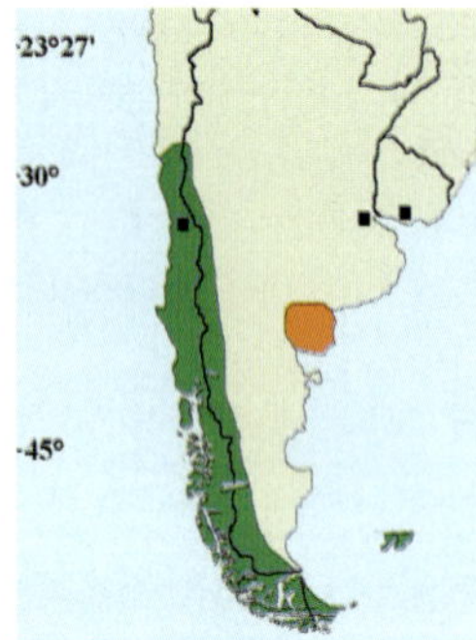

Yahgan story. When we were children, they told us we should not cherish any piece of wood, stone or anything like that, because it could turn into a person. To this girl, it happened. She picked up a beautiful stone she had found in *Tashuani* on Hoste Island at the shore of the sea. And she said to it, "This is my baby." She cared for it until it transformed into her baby. A little head grew, as well as hands and human feet; his body was made out of stone. She had it always with her, took it to her breast, until it bit a chunk from her. So the father of the girl said to her, "What we shall do is throw it into the water because it is bad. We cannot take it with us." So they threw it into the sea.

They thought themselves lucky that he had died already, and they came to a place where they set up a hut, and at night he arrived again, always following his mother. Then, the little stone grew up, and when a man, he caught the women who crossed the canal in canoe, and he killed the men that opposed him. The others ran away. He had many women; it was here at *Lum*. Up in the mountains of *Umushupuliak*, the stone giant tore out big *dihueñe* trees with their roots, carried them on his shoulder for the women to eat.

One day on the mountain, he got a thorn in his foot, and later he got ill. He said, "Oh, my foot is hurting. Take out the thorn!" He lay down in the meadow, and the women pierced into the spot with their *ami* where the thorn was. They hurt him, and he cried out, but it was such a nice, sunny day that he slept and kept on sleeping. The women decided to burn him. With the same trees that he had brought, they surrounded him. The branches were dry, and the fire blazed. His head, hands and feet burned, and his body, which was of rock, began to perish. Small stones jumped out of the fire, and the women put them back into their place with sticks. The women that had touched these stones turned into birds, into Black-Chinned Siskins, or *twin*. Siskins are green and yellow in color, and you can see them in September. When all was burned, they flew away singing their song, joyful to have been liberated from the giant.

This rock that was the giant is still in *Lum*, and when you make a fire near to that place, little stones begin to explode all around.

Yahgan: **Tashúr,** Tachurj, Tōšur
Mapudungun: Chuchan
Spanish: **Cometocino**, Comesebo patagónico
English: **Patagonian Sierrafinch,** Sierra-Finch
Scientific: *Phrygilus patagonicus*

With its eye-catching yellow breast and blue-grayish head and wings, the Patagonian Sierra Finch inhabits the whole range of *Nothofagus* forests in southern Chile and Argentina. This colorful bird usually perches in the canopy of the trees, and one often hears the territorial song of a solitary male coming from high in the branches. At other times it is possible to distinguish the dry "*chic-chic-chic*" of flocks keeping track of one another as they disperse among the foliage of trees or bushes. It inhabits the interior and margins of forests, where it feeds on a great variety of seeds, fruits, insects, and even robs nectar, pecking a little hole at the base of flowers, such as fuchsia or *chillko* (*Fuchsia magellanica*), firebush or *notrü* (*Embothrium coccineum*) and the vine *kolkopiw* (*Philesia magellanica*).

On Chiloé Island, the Patagonian Sierra Finch or *kichan* is considered to be a messenger who brings good news when its "*chic-chic-chic*" sounds, like the striking of two stones together. It brings bad news when it comes near the houses and sings its penetrating and insisting whistle "*chuif-chío, chuif-chío, chuif-chío.*"

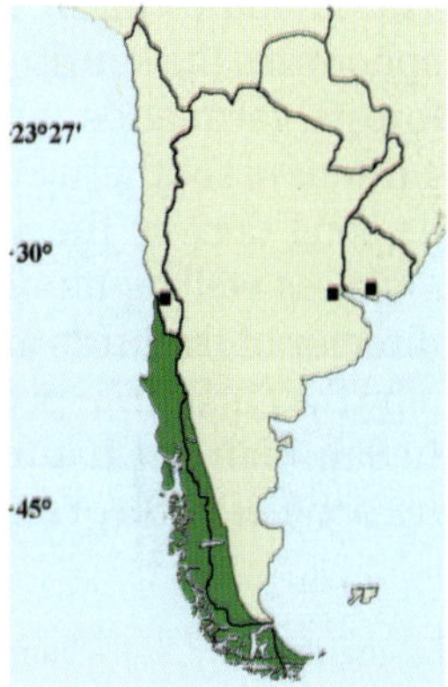

One of the most abundant permanent resident birds of the temperate rainforests between Chiloé Island and Cape Horn, the Patagonian Sierra Finch gives color and life to the months of snow. Perhaps for this reason, the Yahgan grandmothers Úrsula and Cristina Calderón consider the Patagonian Sierra Finch, or *tashúr*, one of the most pleasing birds.

Yahgan: **Hakasir**, Hakásij, Hakásirj, Xakacir
Mapudungun: **Wilki**, Huilqui, Huilque
Spanish: **Zorzal**, Zorzal patagónico
English: **Austral Thrush**
Scientific: *Turdus falcklandii*

The melodic call of the Austral Thrush epitomizes the sunrises and sunsets of the forests, farmlands, and even plazas and gardens of southern South America. From its song derives the *Mapudungun* name *wilki*, as well as its nickname "orchestra director of the birds and nature," given to it by the *Lafkenche* poet Lorenzo Aillapan. In San Juan de Chadmo on Chiloé Island, its song is interpreted as "blessed and praised be God." For the *Lafkenche* poet the melodious *wilki* captures the essence of the Earth, vitalizing all the living beings listening to its whistle:

Singing I say: life – happiness – love. Among songs and dances, through the air to the ear, I say hello to the human, likewise to those who care for the children, to those who are living thanks to virgin Nature. *Chülle Mapu*, earthly paradise, from the mountains to the sea From north to south, all things rejuvenate, returning to spirituality...*	*Pin ñi ülkantun: mongen, ayüwün, poyen.* *Rangiñ ülkantu ka pürün kürüf mew wefi pilun püle,* *Chaliwün che feyti petu penielu fey ñi pichike pu yall,* *Feyti petu mongelelu puche, we mongen lelfün füla.* *Chülle mapu dulliñ mülewe rumel pirentu winkul ka lafken püle* *Pikum püle ka willi kürüf rüf mongetuy wüñotulu püllüyem...*

The *wilki* is the largest flying Passeriform in the *Nothofagus* forests, where it occupies all type of habitats in the interior and edges of forests. They consume all kind of food from the soil and canopy. The *wilki* is so abundant throughout the *Mapuche* territory, that it gives its names to places in the northern, central, and southern regions. For example, in the Cachapoal Valley, Huilquío means place of Austral Thrushes (*wilki* = Austral Thrush, *o* = diminutive of *we*, place), in the Andes Cordillera of Talca, Huilquilemu means forest (*lemu*) of the *wilki*, and on Chiloé Island, Huilqueco means river or water (*ko*) of the *wilki*.

In the Yahgan territory the Austral Thrush or *hakasir* also inspires the women and other members of the culture. The Yahgnas observe how this bird feeds and educates its chicks in trees and along the rivers that

spill through the forests. The Yahgan grandmothers Ursula and Cristina Calderón explain how, when the Austral Thrush has chicks, the mother teaches the oldest one. Grandmother Ursula describes how the "mother *hakasir* rises to a stick. I saw her talking, whistling, teaching her child, constantly moving around the canopy. I was lying down, watching while the *hakasir* told him, 'when you get a brother, you'll have to educate him, and stimulate him to work and wash himself.' In that way the *hakasir* grows. If the Austral Thrush's son doesn't listen to his mother, if the little *hakasir* doesn't obey, she abandons him. The young one stays alone, and doesn't have anything to eat. Then, it feeds on bad *dihueñes* (the fungus *Cyttaria sp.*), those that are overripe, that do harm, and the little *hakasir* dies. If the little *hakasir* obeys his mother, though, he grows up healthy eating good fruit like *amai* (*Gaultheria mucronata*) and *mema* (another fungus species of the genus *Cyttaria*)."

Thus, the Austral Thrush, *wilki* or *hákasir*, expresses to us how the lives and languages of the birds are interwoven with those of human beings in the forested territories of southern South America, as they should be in the rest of the world.

Yahgan story. When the Austral Thrush has chicks, the mother teaches the oldest one. She rises to a stick. I saw the little bird, teaching, talking to her child. I was lying down, watching her. She told him, "When you'll have a brother, you'll have to help raise him, stimulate him to work, to wash himself. In that way the little bird was singing. They flew to the river, where they dove into the water, washing themselves, shaking themselves, and flying back to the tree. In that way they grew up. If the little Austral Thrush doesn't listen to his mother, if he doesn't obey, she abandons him. The young one stays alone, doesn't have anything to eat, lives on bad *dihueñes*, those that are already overripe, that do harm, and he dies. But on the other hand, if he obeys his mother, he grows up healthy living on good fruit, like chaura (*amai or Gaultheria mucronata*) and *mema* (another fungus species of the genus *Cyttaria*).

Mapuche story. With its call the austral thrush typifies the sunrises of the forests throughout the south of Chile and Argentina. The Mapudungun name *wilki* is onomatopoeic from its melodic whistle. To its beat grow the children, and the mothers nurse the babies singing, "*moyoy ta puñeñ, with! lüchütuy püñeñ, will*!"

Mapudungun version. *Tachi wilki pu aliwen koniñi wakeñ ülkantun mawidaantu meu willi püle Chile mapa ka Argentina. Uytukun wilki fey tañi wünül ülkantun wüykeñkantu wünül tremün pichike domo ka feyti pu ñuke moyolu tañi pu püñeñ ülkantuy moyoy ta püñeñ – wüy/wüy ta püñeñ – lüchituy ta püñeñ.*

*From the poem *Wilki, Twenty Winged Poems (Aillapan & Rozzi, 2001)*

Yahgan: **Chílij,** Apuščilix, Hulušénuwa, Apuštöčix, Ūlušönūwa
Mapudungun: **Chedkan**, Chircan, Chedquen, Chedüf, Chüdüf, Chedkeñ, Charkan, Chercan
Spanish: **Chercán,** Ratona común, Raspatortillas
English: **House Wren**
Scientific: *Troglodytes aedon*

The House Wren has a very wide distribution, stretching from the forests of Canada to those of Cape Horn—although this range could include several subspecies or even cryptic species. This little Passerine occupies a great variety of habitat types. In the forests, it nests in trunk cavities. Although the holes have very narrow openings, inside they resemble true caverns. From this nesting behavior derives its scientific name *Troglodytes* (troglodyte, which live in caverns) *aedon* (little). This behavior might be also associated to the Yahgan view of the House Wren or *chilij.*

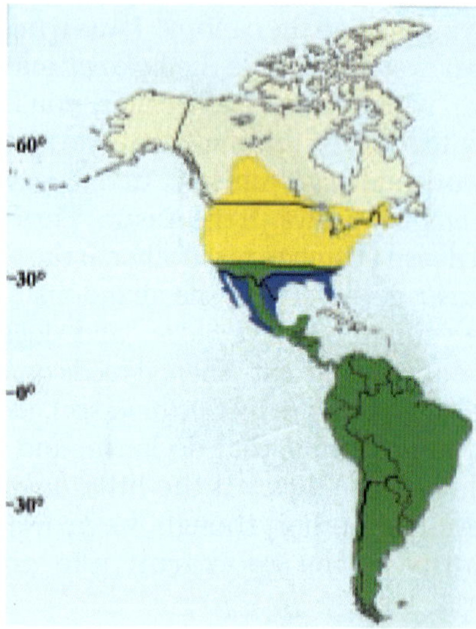

In ancestral times, when birds were still humans, *chilij* possessed numerous wives (*kipa* = woman) that were representative of several bird species that also nest in cavities, like *lasijkipa* (the Swallow-woman) or *tatahurj-kipa* (the Tree-Runner-woman). It is noticeable how *chilij* is constantly hunting insects to feed its chicks, during the reproductive season. *Chilij* has as many as three clutches per year.

The House Wren is also characterized by its melodic vocalizations that include a trilled

song of the male during the reproductive season, as well as numerous raspy vocalizations, such as the territorial call from which its *Mapudungun* name *chedkan* derives. The Spanish name *chercán* derives, in turn, from the *Mapudungun*. In this manner, the onomatopoeia name of the bird is conserved in Chile.

On Chiloé Island, Huilliche populations on the Pacific Coast identify the territorial call of the House Wren, *yec-yec-yec*, with the sound that is produced when "tortillas" are scratched with the shells of mussels or *quilmahue* are ground. For this reason, on Chiloé the House Wren, the same that the noisy Thorn-Tailed Rayadito, is called "raspatortillas" (*Tortilla*-scratcher).

Because the House Wren or *chedkan* sings the whole day, the Lafkenches call gossipy people *charkanas*, i.e., people who pass their time judging the lives and loves of others more than their own lives. Along the coast of Temuco, the *Lafkenche* also call *charkan* a culinary dish that women prepare with *poñü* or potato (*Solanum tuberosum*) and *kollof* or giant kelp (the brown algae, *Macrocystis pyrifera*), because this food acquires the same color as the noisy *chedkan*.

Yahgan: unknown
Mapudungun: **Lantu,** Peutren, Peutrañ
Spanish: **Viudita**, Peutrén
English: **Patagonian Tyrant**
Scientific: *Colorhamphus parvirostris*

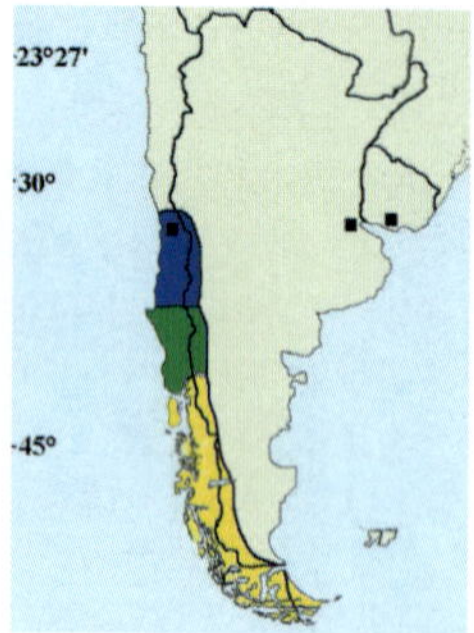

The Patagonian Tyrant is a small insectivorous bird, the single representative of the genus *Colorhamphus*, which is endemic to southern South America. During winter Patagonian Tyrants leave the austral forests, migrating to central Chile or the Neuquén Province in Argentina. In shrublands, forest edges and interior it flies through the lower canopy and perches on the branches of the understory while catching insects. Its small size, and its dark grayish-brown color makes this bird difficult to see. Nevertheless, the tyrant's presence is broadcasted by its long and sad descending whistle *fíuuuuu*...that peals forth from the foliage like a tragic lament. This whistle reminds one of a widow who walks through the forest grieving for the loss of her husband.

In the Hualakura community, where a stone (*kura*) exists in the form of a *huala* or Great Grebe (*Podiceps major*), there lived a widow who with the same tone as the bird lamented the absence of her dead husband, saying, "*layem fel fotrün, layem fel fotrün*" (Oh, how I desire to have my husband together with me!). Other communities of southern Chile think that the name of *viudita*, or little widow, derives from the dark grayish color, which resembles the color of the traditional dark, mourning dress worn in Spanish culture by widows. The name of *viudita* has been incorporated into the modern *Mapudungun* name *lantu* (widow), which complements the traditional name given to this bird, *peutren*.

Mapuche story. The Mapuche bird name *peutren* means knowing the grandparents and ancestors. Chilean people call this bird *viudita*, which means little widow. Perhaps, this name is inspired by the descending pitch of the bird's pitiful call, who sounds like a lady who walks in the forests lamenting her loneliness. In the Nueva Imperial Region, to the southeast of Temuco, on the shores of the Chol-Chol River, a widow lived who lamented the absence of her dead husband every time she got drunk, and recalled the song of the *viudita* in the forests, when she said, "*layem fel fotrün, layem fel fotrün*" (Oh, how I desire to have my husband together with me, Oh, how I desire to have my husband together with me!).

Mapudungun version. *Tachi peutren kimeluwiki kimtukungeyal pu trem epuñpüle lakuwen ka feyti ka che dungun meu lantu pingey feymay ngüman ülkantu yeniel tachi üñüm, chumngechi ngümayawi kiñe domo rangiñ mawidameu lantungüman. Traytrayko waria püle (Nueva Imperial) lafkenkonün antü püle müley Temüko waria inaleufü choll-choll, mülekerkefuy kiñe lantu domo kiñekemeu pütupütungelu fey may rume duamtukefuy ñi fütrayem laludeuma fey pipingey "layem fel fotrü-layem fel fotrü" ka femngechi may konümpamekekefuy tañi ül tachi lantu rangiñ mawidantu meu.*

Yahgan: **Sámakéar**
Mapudungun: Pichpich, Peshkintun
Spanish: **Cachudito**, Cachudito pico negro, Torito
English: **Tufted Tit-Tyrant**
Scientific: *Anairetes parulus*

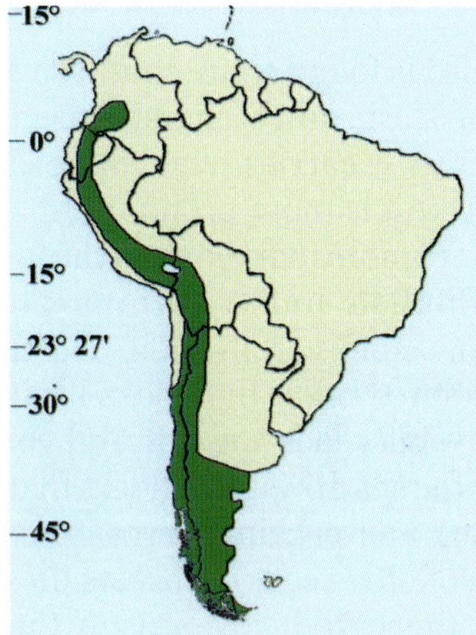

The Tufted Tit-Tyrant gets its English and its Spanish name (*cachudito*) from its long, black crest feathers that resemble a hair-style formerly worn where the hair would be curled at the nape of the neck (*cachito*). Together with the Hummingbird (*Sephanoides sephaniodes*), the Tufted Tit-Tyrant is the smallest bird in the temperate forests of austral South America. The similarity between birds did not pass unnoticed by the Yahgans, who used the name *sámakéar* for both species; the Tufted Tit-Tyrant and the Hummingbird are the only austral birds that weigh less than 10 g.

The Tufted Tit-Tyrant is an anxious bird that moves about constantly between the branches and foliage of trees and bushes, where it feeds on all type of small insects, larvae, and, occasionally, fruits and seeds. From the south of Colombia to Cape Horn, it lives in a wide variety of biomes, utilizing mostly shrubland formations at the edges of forests. Throughout the length of its geographic distribution, it visits human settlements and cities, where it is common in the bushes and trees of gardens and plazas.

The Tufted Tit-Tyrant meticulously builds its nest low in the little branches of the bushes, forming a tiny basket made of grasses, fibers and feathers. From the foliage emerges its trill, which sounds like a smooth chirping in *decrescendo*. It also possesses a territorial song that it usually emits early in the morning that consists of a long sequence of brief, sharp notes.

Yahgan: **Chámuj**, Čamux, Kqaiaminix
Mapudungun: **Chinkol**, Chingol, Mencu, Toque, Mencu Lonco, Meñkutoki, Utruftoki
Spanish: **Chincol**, Chingolo, Copete
English: **Rufous-Collared Sparrow**
Scientific: *Zonotrichia capensis*

This bird is found from southern Mexico to Cape Horn, and is characterized by the grey crest it carries over its head. The male's crest is more pronounced and the colors are more marked than the female's. The English name Rufous-Collared Sparrow comes from their chestnut, or rust, colored collar that looks like a scarf. The juveniles lack the tuft and collar. In winter the sparrows form flocks that move, skipping and pecking, across the soil in search of little seeds or insects. In spring, during the reproductive season, they form pairs and the males frequently emit their melodious and well-known call. They are found in almost all environments, although they are more abundant in shrublands and at the edge of forests.

The *Mapudungun* name *chinkol* means to unite, and denotes the fact that these birds travel in flocks and eat together. This image evokes the collaborative, or shared, work that *Mapuche* people and others carry out on occasions like a *minga* or *mingaco*, where the work is accompanied by good food and fellowship, much like the communal barn raisings that are held traditionally in the United States. Other *Mapudungun* names refer to the characteristic crest that the *chinkol* carries on its head, and that has the form of a hatchet or *toki*. Likewise, he who carries the axe is the war chief or *toki*, and the *Mapuche* nation still aspires to unearth the axe of Caupolicán, their greatest leader: an ax that, it is prophesized, could be found by the *chinkol*. Some names that refer to the *toki* of the *chinkol* are: *utruftoki* (throws the axe; *utruf* = throw), *longkotoki* (with the head of an axe; *longko* = head) and *meñkutoki* (has to do with the "hatchet" or tuft over the brown collar that it has on the nape of its neck; *meñku* = carry over the nape or head).

In the countryside of central Chile the *chinkol* is known as "Uncle Agustín," because its call seems to ask, "have you seen my Uncle Agustín?" On Chiloé Island, it is said that when a *chinkol* perches or sings near the entrance to a house, it announces the arrival of a letter or visit of good will.

Yahgan: **Hashpúl**, Hĕspul, Ŭspul
Mapudungun: **Püdko**, Huelko, Diucon
Spanish: **Diucón,** Huilco, Hurco, Ojos colorados
English: **Fire-Eyed Diucon**
Scientific: *Xolmis pyrope*

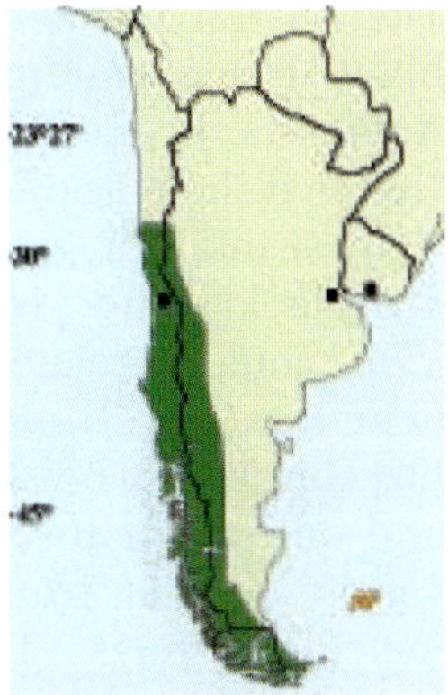

The call of the *püdko* suggests the falling of water drops or the passing of heavy clouds. The *Lafkenche* communities that inhabit the sector of Lake Budi associate this call with the coming of drizzle because as its name indicates, the *püdko* separates the clouds (*püd* = to separate two clouds) loaded with water (*ko*). The Huilliche communities of Chanquín and Huentemó on Chiloé Island believe that when the *püdko* or *huelko* rises and falls over the same branch it announces bad weather. For these communities the *huelko* has spiritual powers and is a good shaman that accompanies the travelers through the forest, looking out for them from the canopy of the trees and bushes. For other Huilliche communities it is a "sent one" that the witches dispatch to test the reactions of a person under observation. For this reason, it is treated with respect.

The Yahgans also say that one has to treat the Fire-Eyed Diucon or *hashpúl* well because it is a powerful shaman. It is wise to take care with this "red-eyed one," as it is also known on Navarino Island. Grandmother Úrsula Calderón warns that one should never throw stones at it the *hashpúl* unless they want strong storms with southern wind or *ilan* to descend upon them:

Yahgan story. My grandfather said that the *hasphúl* or Fire-Eyed Diucon was a signal of bad fate. If we saw him, we were not to throw any stones after him. If we did so, the *hasphúl* would bring rain, thunder and snow. My brothers did not believe my grandfather, and they threw stones after a *hasphúl* in order to find out whether he was right. So soon afterwards the bad weather began; it came with rain and thunder. "Wow!" they said. "It is true what the grandfather told us, never again we will do so." Later, my mother confirmed: when the grandfather says something, we have to obey him, because he knows. And surely, it is bad to throw stones after the *hasphúl* bird.

Yahgan: **Tátapuj,** Kátapuj, Tetapux, Totupux
Mapudungun: **Küreu,** Cureu, Chihuanco, Quereu, Furare, Kürew
Spanish: **Tordo**, Tordo patagónico
English: **Austral Blackbird**
Scientific: *Curaeus curaeus*

The Austral Blackbird is a completely black bird, endemic to Chile and Argentina. In these countries, they say "it's as if the blackbirds were bathed in Chinese ink," for, even the feet and eyes are black. The *Mapudungun* name *küreu* could be related to the color black (*kurü*), as well. Most of the year blackbirds form boisterous flocks inside and outside of forests, which emit a variety of calls, from strident cries to melodious songs. These vocalizations permit the blackbirds to keep their group together while they feed. The *Mapuche* name *küreu* might be onomatopoeic with its most common call. From the *Mapudungun küreu*, the Jesuit abbot and naturalist Ignacio Molina derived the Latin name: *Curaeus curaeus*.

The *Huilliche* people of Chiloé Island attribute distinct meanings to the variable calls of the *küreu*. If they are melodious, or if they cross you in the road and call softly, then your business will go very well, but if its call is harsh your fortune will be adverse. The blackbird possesses an elegant black-blue color that has given him the nickname of "dandy."

Julia González tells that when the Yahgan see the blackbirds or *tátapuj* descend to the coasts of Na-varino Island and the Archipelago of Cape Horn, it is because snow storms will come, or because it has already snowed abundantly in the mountains or *tulára*. For the Yahgan and other inhabitants of the extreme southern region, albino blackbirds that "take the color of the snow" are well known, as well.

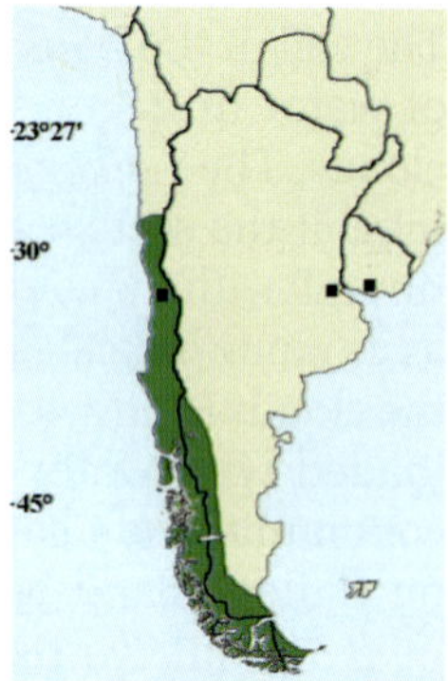

Yahgan: **Sámakéar**, Omora, Samakéa
Mapudungun: **Pinda**, Pinguera, Pinuda, Piñuda, Pigda, Pichi pinda
Spanish: **Picaflor chico, Colibrí,** Picaflor rubí
English: **Green-Backed Firecrown,** Crown Hillstar, Fire Crown
Scientific: *Sephanoides sephaniodes*

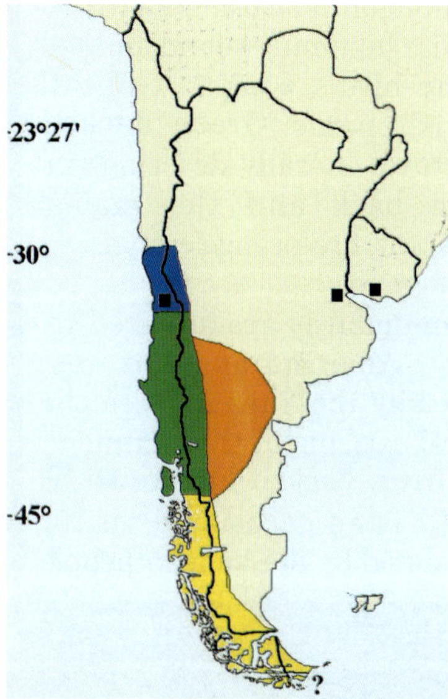

Hummingbirds are a bird family found exclusively in the Americas, whose species are concentrated in the tropics. They dart about, and then suspend themselves, seemingly motionless in the air, beating their nearly-invisible wings up to 200 times per second. This extraordinary type of flight is fueled by one of the highest metabolic rates known for living organisms. Because the climate of the tropics is warm, the energy needed to maintain body temperature and fly is less than at cold, high latitudes. For this reason, it is surprising to find a species of hummingbird that reaches to the southern extreme of South America, in the Chilean Antarctic Province. Moreover, in the austral forests there are, however, more than ten species of plants with red, tubular flowers whose pollination depends on this little bird. As it flies among the flowers, the Green-Backed Firecrown brings an unexpected tropical character to the austral forested landscapes.

The presence of the Green-Backed Firecrown excites not only biologists. Its Spanish name, *picaflor*, means to prick (*picar*) the flowers (*flores*). Its

Mapudungun names *pinda*, *pinguera*, *pigda* or *piñuda* are related to the verb *pigudcun* (to rub together), and denotes the sound the wings make during rapid flight. The English name for hummingbirds also alludes to this peculiar humming sound emitted by the bird's wings. The full English name Green-Backed Firecrown literally describes its green back and the males' iridescent, fire-orange crown.

Hummingbirds are admired by many Amerindian cultures, especially the Yahgans. In the austral extreme of the continent, the Green-Backed Firecrown or *omora* is an occasional visitor considered by the Yahgan people

to be a bird and, at the same time, a small man or spirit who maintains social and ecological order. Today, the figure of *omora* has inspired a conservation initiative that integrates the biological, ecological, anthropological, social and cultural dimensions of life with the goal of promoting the well-being of all species, including humans, in the Cape Horn region. Narratives, such as the following, teach us how *omora* achieves such a life-affirming integration since long ago:

Yahgan story. In ancestral times, when birds were still humans, a great drought occurred in the region of Cape Horn and the people were dying of thirst. Only the wily fox, known as *Cilawáia*, found a lake. He told no one and built, instead, a solid fence around it so that no one could enter. Hidden as such, he drank lots of water, concerned only with himself. However, after some time the other people discovered this lake and went as a group to ask the fox for a little bit of water. *Cilawáia* did not even listen to their supplications and expelled them with rude words. The condition of the people got increasingly worse and when they were at the point of death, they decided in their desperation to send a message to the hummingbird or little *Omora*, an occasional visitor who had saved them before in times of crisis. *Omora* was always prepared to help others and arrived quickly. Although diminutive, this little tiny bird-man-spirit is braver and bolder than any giant. When he arrived, the downcast people told him about their great sufferings, and, indignant, *Omora* raised himself and rapidly undertook the flight towards where the egotist *Cilawáia* was, and he confronted him directly saying: "Listen! What is it that the people are telling me? Have you had access to a lake full of water, but you don't want to share it with other people who soon will die of thirst!" The fox replied, "Why should I worry myself about the others? This lake has only a little water, and it barely is enough for me and my closest family. I cannot give anything to other people because soon I'll suffer thirst myself."

Upon hearing this, *Omora* became furious and without replying to the fox, returned to the settlement. He reflected briefly and rapidly rose again, took his sling and flew back to the fox. Along the way he collected several sharp stones, and when he was within sight of the fox and sufficiently near, he shouted to him: "Are you going to share your water for now and for always with the other people? Come on, don't be so selfish. If you don't spare water, they surely will die of thirst!"

The fox answered indifferently, "They can all die of thirst. Why should I worry? I can't give water to each one of the people who live here, or else my family and I will also die of thirst." *Omora* was so offended that he could not control himself any longer and furiously fired a sharp stone with his sling, killing the fox with his first throw.

The rest of the community had been watching and became happy, running to the place. Rapidly they broke down the fence, went to the lake and drank, satiating their thirst. They drank so much that the lake became completely empty and a few birds that arrived too late barely found a few drops with which they moistened their throats. Then, the wise owl *Sirra* or *Sita*, the grandmother of *Omora*, said to the birds that arrived late, "collect mud from the bottom of the lake and fly to the peaks of the mountains, and fling the mud over them." The little birds flew, and their balls of mud gave birth to springs that became water courses that spouted from the mountains, forming small streams and large rivers that flowed through the ravines. When all the people saw this, they were extremely happy and drank great quantities of fresh, pure water, that was much better than the water of the lake, and now all find themselves safe again. Today all of those water courses still flow down from the mountains and provide exquisite water. Since that time no one has died of thirst.

Yahgan: out of bird´s range
Mapudungun: **Fütra pinda,** Pinguera, Pinuda, Piñuda, Pinda
Spanish: **Picaflor gigante**
English: **Giant Hummingbird**
Scientific: *Patagona gigas*

The Giant Hummingbird or *fütra (big) pinda* is the largest hummingbird of the world. It is surprising to see it beating its wings and emitting its short, sharp whistles in the middle of the forest. Like the Firecrown, it migrates in winter to central Chile, but some migrate even further north, since the Giant Hummingbird is distributed through the Andes up to Ecuador. Many rural folks in southern Chile think that during winter some hummingbirds go into tree cavities and hibernate, to return to the forests with renewed vigor when the flowers bloom in spring. Regarding the hummingbird, the poet Lorenzo Aillapan sings:

It nourishes itself with the morning dew, the honey and the liquor of the flowers... Tiny, little bird with so much speed, that while flying, its wings cannot be seen. It sleeps, or nods off, in April, and awakes, or revives, in October, when there are many flowers.	*mongekey mülüngmew ka modkoñomew* *ka llumed pezdkiñmew* *kürüf trüri ñi miyawün pichi üñüm* *üpünüyüm pefalkelay ñi müpü* *Umagtukey peuwün küyen püle* *Nepetu mongetukey tromüngen küyem* *püle**

At night, hummingbirds fall into a state of torpor, and during winter some individuals might seek refuge in holes of trunks and other protected sites for a prolonged torpor period. Among all animal species, these marvelous little birds present the highest metabolic rates, and they still open many new questions to science.

Hummingbirds not only excite scientist and poets, but also women who want to have children. On Chiloé Island when a woman cannot maintain her pregnancy, it is said that she should trap a hummingbird, caress it in her hands and liberate it immediately. That way, she will receive the powers of fertility she needs to be a mother.

Several places on Chiloé have *Huilliche* names that allude to hummingbirds: *Pindapulli*, in the mountains of Pichué, means mountain (*pulli*) of the hummingbirds (*pinda*); *Pindaco*, water (*co*) of hummingbirds, indicates a leafy place with vines where Chonchi Creek passes; *Pindal*, place of many *pindas* or hummingbirds, lies just to the north of Chonchi. The Giant Hummingbird has left its mark on many places in the *Mapuche* territory of southern Chile, as well as on aboriginal cultures farther north; it is the same species that inspired the giant hummingbird geoglyph at Nazca, Peru.

**From the poem Pinda, Twenty Winged Poems (Aillapan & Rozzi, 2001)*

Yahgan: out of bird´s range
Mapudungun: **Diuka,** Shiwka, Fiuca, Diwka, Viuca, Diuca
Spanish: **Diuca**
English: **Common Diuca Finch**
Scientific: *Diuca diuca*

Pewü mew puliwen sollpiwükey diwka! On spring (*pewü*) early mornings (*puliwen*), the Common Diuca Finch (*diwka*) sings (*sollpiwükey*) this lovely song). Considered by many bird-watchers to be the loveliest bird-song of Chile, the Diuca Finch beautifies the natural environments and countryside of southern Chile and Argentina with its melodic successions of early-morning whistles: *diu-diu-diu-diu*. This song inspires the onomatopoeic *Mapudungun* name, *diwka*, and initiates the morning concert of birds in springtime. With admirable precision the song announces the hour of the dawn which is why when the country people wake up early, they often say they "arise with the first diuca call." When the diuca sings, it is easily visible: usually perching on the canopy of a tree, a protruding branch, a post or a fence wire. It has a characteristic white throat that contrasts with its predominantly grey color. It is distinguished from the Fire-Eyed Diucon (*Xolmis pyrope*) by its black eyes.

The Diuca Finch prefers to inhabit prairies, shrublands and other open habitats, but it can also be found on the edges of forests, a trait that the town Diucalemu is named after. Diucalemu means forest (*lemu*) of the *diucas*.

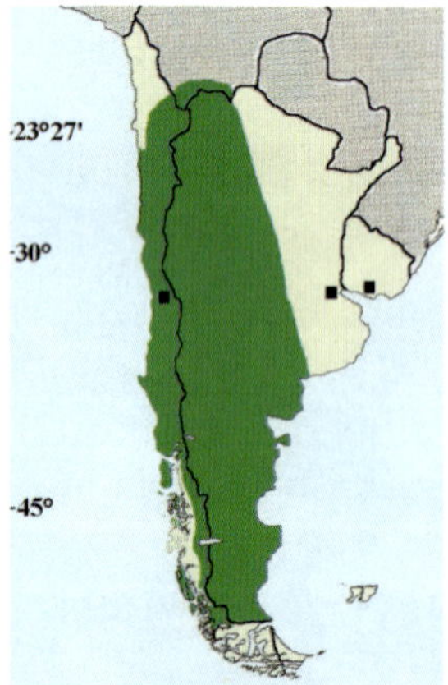

The adult *diucas* feed on seeds, and occasionally berries and other fruits from the bushes and forest. The young feed, however, on a variety of insect and invertebrates. Because it eats mostly seeds as an adult, the Diuca Finch is characteristically called a granivore, yet is really an omnivore.

Yahgan: **Wichóa**, Wičōa
Mapudungun: **Lloyka,** Loica, Lloica, Loyka
Spanish: **Loica**, Pecho colorado, Loica común
English: **Long-Tailed Meadowlark,** Red-Breasted Starling
Scientific: *Sturnella loyca*

The poet Lorenzo Aillapan says that the *Mapudungun* name *lloyka* comes from the terms *llako* (to heal) and *lawen* (remedy). According to this *Mapuche* Bird-Man, the *lloyka* is a healer-bird that, like the *machi* (Mapuche shaman), gets its medicines from the plants of the forests. Perhaps this *uñüm* (bird) still desires to heal its chest, which carries the blood of the Mapuche nation and the flower of the *kolkopiw* (*Lapageria rosea*).

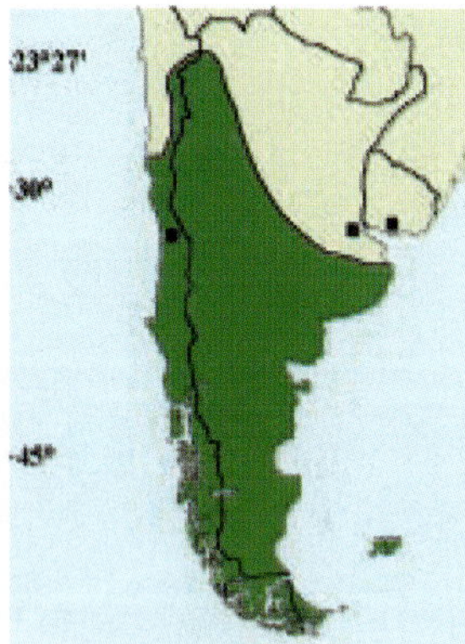

Have no fear Great Father and Great Mother, I will have the remedy in time, abundant and good. For this, I call myself another red-breasted being. I have always healed my people with pure herbs. For this, my name is *full-time healer*.	*Llükalayaymi turpu fütra chaw ka eymi kude ñuke* *Inche ta fey lawentuayu feula fachante frentren ka tutelu* *fey mew ta llapüm chefe kelü pütra pingeken* *Ka fey rumel llakotukey pu che re mapu lawen mütem* *Inche ta fey mew mongelchefe rumel lloyka pingen**

In central Chile it is said that its call expresses, "with a knife it was, with a knife it was," alluding to the moment when the Long-Tailed Meadowlark thrust a knife in its chest, which shines as if still bleeding, covered in red.

In the Yahgan territory at the austral tip of the continent, Long-Tailed Meadowlarks are abundant only along the northern coastline of the Beagle Channel. South of that, they are very scarce. The grandmothers Úrsula and Cristina Calderón remembered the Long-Tailed Meadowlarks or Red-Breasted Starlings from their young years when they lived on the lands of the Anglican missionaries at Harberton Ranch in southern Tierra del Fuego, Argentina. However, after so much time and being infused with other notions of the bird, it took more than one year, working in the field together, before Grandmother Úrsula could remember the Yahgan name, *wichóa*. May be this healer bird, *wichóa* or *lloyka*, can heal our memory at the southern tip of South America, and communicate to us the importance of knowing, watching over and valuing the diverse bird species, and appreciating the diverse cultures that inhabit the austral forests.

Mapuche story. The *lloyka* or Long-Tailed Meadowlark is a healer bird who transports medicines from one place to another. Among the medicinal plants the *lloyka* transports are pieces of a sacred tree named *foye* (*Drimys winteri*). These pieces are applied to the *rehue*, carved tree trunk posted outside the house of the shaman, who is known as the *machi*. The *rehue* symbolizes the powers of the shaman, and like her house or *ruka*, faces the northeast where the sun or *antü* rises in the morning.

Mapudungun version.
"¡Wichin wichin wichin wichin wichin küpalu piam pu lonko Trapuuu
wichin wichin küpalu piam lantuuu kangey trekan lantuuu!"
Tachi lloyka llakonchefe üñüm yekey fill lawen fill püle. Tachi lawen foye pingelu lloyka ta yey muñküpüle fey wefi foye (Drimys winteri). Allangechi lawen aliwen fey kuyfimel rewengekey, kiñe mamüll foye feyti papay machi kintuy tukuy tañi rewe wülngiñ ruka püle, cheutañi tripan antü. machi, dulliñ, lawentufe domo tuwükülelu wüne newen püllümeu llapümkelu kutran kafeyti fill weda neyen kutran).

**From the poem Lloykakawün, Twenty Winged Poems (Aillapan & Rozzi, 2001)*

Yahgan: out of bird´s range
Mapudungun: **Kamtrü,** Rara, Cadi
Spanish: **Rara**
English: **Rufous-Tailed Plantcutter**
Scientific: *Phytotoma rara*

Among the birds inhabiting the temperate forests of South America, the Rufous-Tailed Plantcutter is the only species that is primarily herbivorous. There are very few small herbivorous bird species in the world. Most herbivorous birds weigh more than 1 kg and browse on grasslands, such as ostriches and geese. The Rufous-Tailed Plantcutter, in contrast, flies among trees and shrubs, eating the flowers, buds and tender leaves of several plant species with its finely serrated bill.

The male has a distinctive copper color under tail coverlet, and on the belly, throat, and crown. Females are duller, but both sexes have bright red eyes. Finally, the most peculiar characteristic of this bird is the unmistakable territorial call that the male utters while perched on a high tree branch. This call consists of a sequence of rasping sounds that Chilean ornithologist Guillermo Egli compares with the sound of a matraca rattle.

It is precisely from this strange, percussive call of the Rufous-Tailed Plantcutter (*ke-ke-ke-ke-kreeé* o *rara-rara-rara*) that the onomatopoeic *Mapudungun* names *kamtrü* and *rara* derive. This call is heard as much in the forests of the mountainous *Pehuenche* territory, as it is on the coasts of the *Lafkenche* territory. *Kamtrü* lives not only in the forests or *mawida*, but also in the agricultural lands or *ngan*. Lorenzo Aillapan explains that, on the coast of Temuco, the Rufous-Tailed Plantcutter has a special preference for broad bean plantations. For this reason this bird is locally known as *cortaplantas* (plantcutter). However, due to its low numbers, *kamtrü* are not harmful for crops, and as a result its hunting and capture are prohibited.

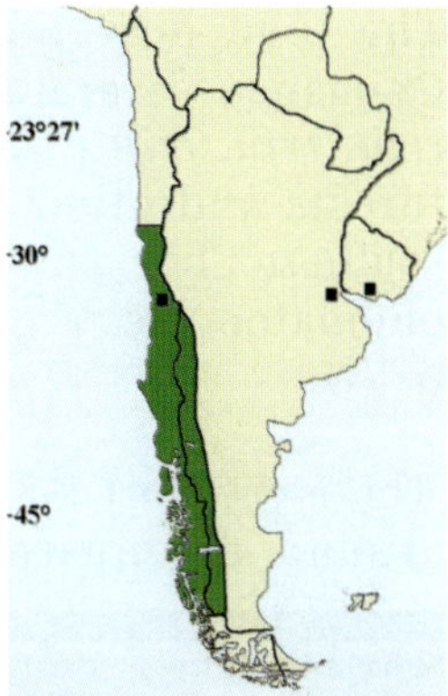

Yahgan: **Yoskalía,** Yookalía, Ȳoakölia
Mapudungun: **Triwkü,** Triuki, Tiuque, Chiuque, Triuquem, Chiwkü
Spanish: **Tiuque,** Chimango
English: **Chimango Caracara,** Tiuque
Scientific: *Milvago chimango*

The loud descending cries of the Chimango Caracara (*triiiiuuuu, triu, triu, triu, triu ....*) characterize the forests and other environments of southern Chile, and is the source of its onomatopoeic *Mapudungun* name: *triuki*.

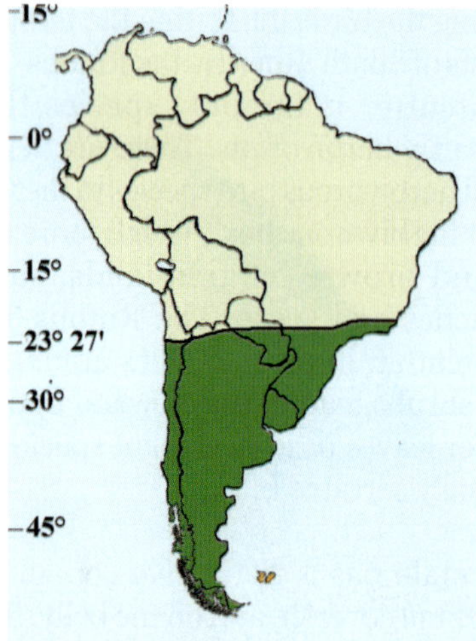

This is the most common raptor in and around the austral forests. The Chimango Caracara uses trees to sleep and nest in, and builds its large nests with twigs and branches. Although it is mainly a scavenger, it is omnivorous, as well, hunting frogs, lizards, mice, small fish, insects, earthworms, larvae, caterpillars and even slugs. Therefore, it is a very beneficial bird for agriculture. When cows are browsing in the prairies or people are hoeing the land, groups of up to hundred Chimango Caracara are seen eating insects on the ground.

The *Huilliche* on Chiloé Island consider the Chimango Caracara or *triuki* a "suspicious bird" because witches use them and even transform themselves into them. When the Chimango Caracara lands on the roof of a house it can be a witch that listens to the conversations of the people inside.

For the *Lafkenche*, in contrast, the *triuki* brings happiness by accompanying them during agricultural labors and by removing the lice from the animals. In addition, they help the farmers because their cries call rain to soak the land, and they eat grubs and other crop pests.

Mapuche story. The cry of the *triukü*, *triiiiuuuu, triu, triu, triu, triu...* invokes water, humidity and rain, because they feed on worms and grubs in the moist earth. Consequently, it is a beneficial bird for farmers.

Mapudungun version. *Feytachi pin dungun triuki trüüuuuu, triu, triu, triu triu, mütrümi komawün, ka fochon ka feyti mawün, feykam ikelu düllwi, katachiwera piru fochon tuwe meu. Feymeu tati, küme üñüm kellukefi pu küdaufe fünketrankelu.*

Yahgan: **Ketéla**
Mapudungun: **Traru,** Traro, Taru
Spanish: **Traro**, **Carancho,** Huarro, Caracara
English: **Southern Crested Caracara**
Scientific: *Caracara plancus*

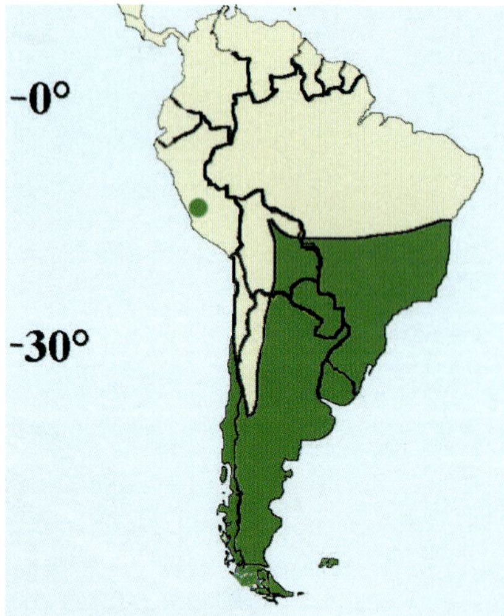

The presence of the Crested Caracara is easily detected by its loud squawk, *K-a-r-a-k, K-a-r-a-k*, or *traru-traru-traru...*, which is heard from long distances away. These calls are echoed by Mapuche name of *traru*, as well as the Spanish *traro* and English Caracara. The latter appears in its scientific name (*Caracara plancus*), and is also used as one of the common Spanish names. When it calls, the caracara makes a typical movement of the head, throwing it back.

Given the prominent crest that it carries over its head, the *Lafkenche* people compare the *traru* with the Chilean

national police, which are figuratively called *longko-traru*, since that like a *traru* they wear a cap on their heads (*longko*).

The Crested Caracara is a typical bird of prey and scavenger in the austral forests and other environments of South America. Only five years ago, scientists recognized it as a different species from its sister species, the Caracara of Central and North America (*Caracara cheriway*). Both species are characterized by their distinctive vocalizations and by the black crests over their heads. The Crested Caracara builds extraordinarily large nests in the canopy of tall, old trees, which are platforms of twigs and branches covered with soft material. For this reason, conservation of the forests is very important for the survival of this species. Crested Caracara are year-round residents, and pairs will stay on the same territory for many years. Young birds have to travel many kilometers to establish their new territories. Omnivores, they eat small mammals, lizards, frogs, insects, crabs, worms and carrion. Occasionally, they attack lambs. The latter image should not diminish the positive contributions that Crested Caracaras makes by "cleaning" the countryside of pests and controlling plagues.

The *Mapuche* people have great admiration for raptors, and the *traru* has a special historical significance to their culture. The lineage of eagles and Caracaras gave rise to the greatest of all *Mapuche* warriors: *Leftraru*, whose name means quick (*lef*) *traru*. Known by the Spaniards as Lautaro (1535-1557), *Leftraru* was a superb war strategist organizing his people into distinct flanks of attack in a way that startled the Spanish, *Leftraru's* martial skills were complemented by his profound knowledge of the topography of his land and by his ability to forecast climatic phenomena. This deep understanding of the *Mapuche* territory allowed him to defeat resist the Spanish conquistador Pedro de Valdivia. The war strategies of this valiant and intelligent warrior have inspired not only generations of *Mapuche* people; Poet Lorenzo Aillapan points out that Napoleon was also inspired by the tactics of *Leftraru* that confused the enemy and attacked him simultaneously from different fronts.

The name of Lautaro's father was *Kürüñamku*, which means black (*kürü*) eagle (*ñamku*). As is the case of many *Mapuche* families' ancestry, the lineage of this hero's family is intimately linked to the swift raptor birds or *norche*: people (*che*) who love justice (*nor* = rights) and liberty (*naytun* = liberty). Unfortunately, the clear-cutting of native trees in the Lafkenche territory has eliminated the places where this noble bird usually nests, and its populations are diminishing drastically. Our goal is to contribute to the protection of these coastal, native forests and to preserve the Crested Caracara and the values inherited from *Leftraru.*

Mapuche story. *Leftraru* means quick *traru* or Caracara. This was the Mapudungun name of the great Mapuche warrior-hero, known as Lautaro, whose father was called *Kürüñamku*, which means black eagle. So, the family lineage of this hero, like that of many Mapuche families, is intimately tied to birds.

Mapudungun version. *Tachi Leftraru feyta wüne fütra aukafe mapuche ngey, kimngetuy Lautaro üyngen, ka feytañi chau Kürüñamku pingefuy (águila negra). Femngechi, kom küpal lakuy wüne ñidolaukafe, kafey inapalu pu mapuche, üñüm meu tuwükuley laku küpal.*

Yahgan: **Ǩikinteka**, Aǩimakaia
Mapudungun: **Okori,** Ocori, Lleullequen, Lilpillañ, Lleylleykeñ
Spanish: **Cernícalo,** Halconcito común, Quechi-Quechi
English: **American Kestrel,** Sparrow Hawk
Scientific: *Falco sparverius*

The American Kestrel is a species of small falcon that is abundant throughout the Americas, with the single exception of the Amazon rainforests. In the forests of southern South America, the *Mapudungun* names given to this little falcon suggest the three cognitive dimensions from which the *Mapuche* culture names the birds: 1) onomatopoeia, 2) cultural-spiritual, and 3) natural history:

1) The name *okori* describes the sound of the whistle the kestrel makes when it imitates the hiss of its snake or reptile prey. Among the Mapuche, the word *mutrir* indicates the most intense hiss of a snake or *filu*, which is also the warning cry that one calls to another. This hiss can be harmful for human beings.

2) The name *lilpillañ* expresses the cosmology in which the kestrel and other raptors acquire the character of guardian (*pillañ*) of the hills and volcanoes with their cliffs and rocks (*lil*). The American Kestrel or *lilpillañ* inhabits the mountain slopes, occupying a variety of environments. *Lilpillañ* is also a skillful flyer, and during the flights makes a series of strong, short cries, which give it the character of a protector bird. In the mountains it nests in all types of natural cavities of rocks, cliffs and in the austral forests its takes advantage of the hollows excavated by woodpeckers in large trees.

3) The name *Lleylleykeñ* comes from a natural history observation and refers to the kestrel's habit of suspending itself in the air to observe its prey. The stationary-soaring kestrel concentrates its senses, listening to sounds and using its intensely perceptive eyes to spot the slightest movement of prey below them. This disposition is similar to a state of being quiet and concentrated with alert senses, which the Mapuche call the *lleylle* state. For this reason, it is called *lleylle*; the suffix *keñ* is applied the name a noun, which describes how the bird behaves. *Lleylleykeñ* performs its stationary flight mainly on open lands where it hunts rodents, little birds, reptiles, and a large variety of insects.

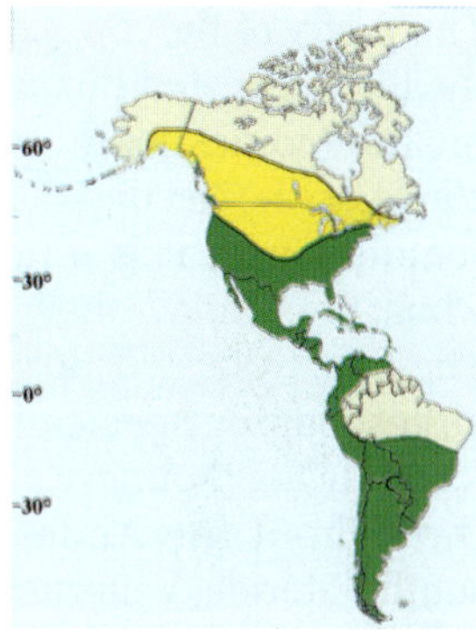

Yahgan: **Kíkintéka,** Kíkentéka
Mapudungun: **Kalkiñ**, Caltchi, Calquin
Spanish: **Águila**
English: **Black- Chested Buzzard-Eagle,** Shield Chested Eagle
Scientific: *Geranoaetus melanoleucus*

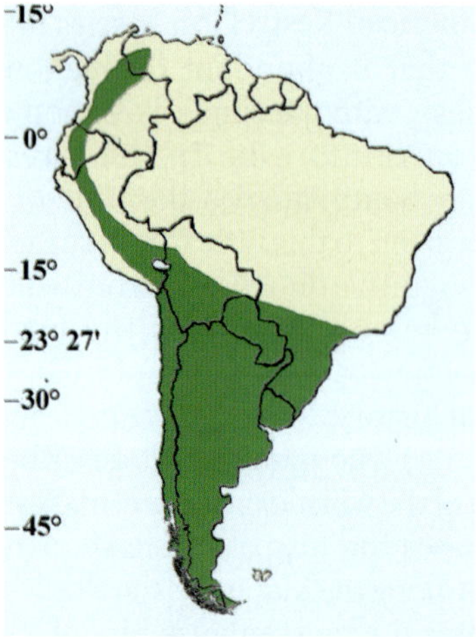

The prefix of the *Mapudungun* name for the Black-Chested Buzzard-Eagle (*kalkiñ*) means "comes from another place." It alludes to the fact that this bird comes from remote lands; it is a notable visitor for whom the *Mapuche* have great admiration. The Black-Chested Buzzard-Eagle is the largest raptor species in the temperate forests of South America. Its range extends throughout the Andes: from north to south—Mérida, Venezuela to Cape Horn—and from east to west—the forest south of the Chaco in Paraguay and the xeric shrublands of Córdoba, Argentina to the Pacific Ocean.

The Black-Chested Buzzard-Eagle is seen from far away flying in high altitude circles and, from close at hand, perching over a tall tree at the edge of the forest. It builds its nest of sticks over the canopy of tall trees or on rocky cliffs, and it occupies this nest for several years. It feeds on a variety of medium-sized prey, including rabbits, hares, rats and also, in the extreme south, the exotic, invasive beavers and muskrats. Therefore, it is a beneficial soecies to controlling the numbers of rodent and small mammal populations.

In flight, the Black-Chested Buzzard-Eagle presents an unmistakable triangular shape; the base of its wings is very wide and the tail has a wedge shape. Its upper parts are grey and black, in contrast to the underside, which is white, except for the characteristic "black shield" on its chest that provides the source of its English names, Black-Chested Buzzard-Eagle or Shield Chested Eagle.

In the ancient stories recorded in the territory of the *Pikunche*, the white abdominal feathers of the eagle are of central importance. In the area of Vilches, towards the mountains near Talca, it is believed that in ancestral times, the eagle saved humanity. When there was a great downpour caused by *Kai-kai vilu*, the people went up the mountain where *ten-ten vilu* lived. As the flood waters rose, the people climbed higher and higher up the mountain. When they climbed as high as they could go and were still unable to escape the deluge, the eagle appeared from the heights of the Andes and with one flick of its beak opened a crater in the volcano. There, the ancestors of *Mapuche* took refuge. Until this very day, the volcano remains covered with snow, the feathers of this eagle who saved the *Mapuche* people.*

**The story of* kai-kai *and* ten-ten *presents many variations in different regions of the Mapuche territory, and at different historical times. The version presented here was recorded in the area of Altos de Vilches (VII Region, Chile) in 1994.*

Yahgan: unknown
Mapudungun: **Ñamku,** Ñancu, Nangu
Spanish: **Aguilucho**, Aguilucho común
English: **Red-Backed Hawk**
Scientific: *Buteo polyosoma*

The *Mapudungun* name of the Red-Backed Hawk means "departed spirit" (*ñam*) which returns eventually to loved ones, and is of good omen. The suffix *ku* makes the name a noun. The Red-Backed Hawk flies high above the forests of South America, detecting its prey with sharp eyes in order to launch itself and capture them.

Its plumage is found in many varieties, but the characteristic plumage of the adult can be recognized even when it flies high overhead; the white colors on the front and the black edges of the wings and tail always stand out. It usually perches on rocks in the mountains and exterior branches of large trees in the austral forests.

In the south it constructs nests from branches on tall trees, from where its loud, piercing cry can be heard. It feeds on rodents and birds, but also consumes reptiles like the snake and lizard, and occasionally even insects such as beetles. Inhabiting all types of environments, including arid zones and rainforests and from sea-level to 4,500 m, it is found

throughout the Andes from the Sierra de Santa Marta (Colombia) to Cape Horn (Chile).

Within the *Mapuche* territory many people liked the name of this bird and called themselves *Ñamkupil*, which means "in the spirit (*pil* or *püllü*) of the Red-Backed Hawk (*ñamku*)"; or, *Ñamkulef* which means "swift (*lef*) *ñamku*"; or *Ñamkuan* which means "*ñamku* of the sun (*an* or *antü*)"; and, *Ñamkucheo* which indicates that the person has a great likeness (*cheo*) to the *ñamku* or Red-Backed Hawk. In addition, for the Mapuche people, *Ñamku* is the guardian of the herds rebanos, and is respected and always welcomed. The deep empathy between the people and *ñamku* is expressed not only in names and habits, but also in ancestral stories, such as the following:

Mapuche story. Many people carry a feather of the *ñamku* as a keepsake so that things always go well. Sometimes, the *ñamku* removes a feather and allows it to fall. This happened to the young man, *Mankian*, who, after collecting the feather that the *ñamku* that had fall, became the Sea Guardian, called *Dumpall* or Guardian of the West. *Dumpall* accompanies the Volcano Guardian of the Southeast named *Pillán*, the Princess of the Sun from the Northeast named *Anchümallen*, and the Great Visitor who has the treasures from the North, *Witranalwe*.

Mapudungun version. *Fentren che yeniey ñamku pichuñ rumel konumpa nieyal fey ñi kümelkayawual. Ñamku ta ütrüfnagomi kiñe pichuñ fey ta tuy tachi weche Mankian, feyta ngenlafkengepuy Dumpall pingetuy. Tachi ngen lafken konün antu püle, kiñeuküley tachi Pillan engu willi kürüf püledegüñ tuwülu, ka feyti Anchümallen tripan antu tuwülu, ka tachi Witranalwe, pikun kürüf püle tuwülu futra ülmen kupal.*

Yahgan: **Kulalapij,** Kilakimõs, Cazapollo
Mapudungun: **Kokoriñ,** Peuco, Peucu, Kokori
Spanish: **Peuco,** Gavilán mixto
English: **Bay-Winged Hawk, Harris´s Hawk**
Scientific: *Parabuteo unicinctus*

In the rainforests of the *Mapuche* territory in the Andes Cordillera that rises above the Malleco River, the name of the village of Melipeuco denotes the place of encounter of four (*meli*) Bay-Winged Hawks (*peuco*). This raptor also receives the *Mapudungun* name of *kokoriñ* and is considered to be a bird of astute and sharp perception. It has strength, good sight and even better hearing. In the area of Chillán when the mothers do not the suitors of their daughters, they refer to him as the *peuco*. It is common to hear them say, "I don't like this *peuco* one little bit." On the other hand, the men refer to the "*peucas*" as the girls of ill repute.

The Harris's Hawk is a large dark brown bird of prey, whose territory extends from the southern United States to southern Chile and Argentina. It is quite frequent down to the area of Aysen and Calafate, but is occasionally found further south.* It usually perches among the branches of trees or bushes, as well as on fence posts in the country. It preys upon hares, rabbits, rodents,

reptiles and many types of birds, including barnyard fowl—which leads many people in the countryside of southern Chile to call the Harris's Hawk "*come pollo*," or chicken-eater. In the countryside it is frequently hunted and persecuted, because it is said: *kokoriñ rume iñfitukey achawall mew* ("the Harris's Hawk is damaging because it eats the chicken"). Underscoring its reputation for capturing hens and chicks in the farm yards, the poet Aillapan calls this hawk *weñefe üñüm* (chicken thief-bird). Likewise, in the valley of the Ñuble River, the children have a game that mimics the hawk's mannerisms. The boys and girls act like the hawk and hen, chasing each other. As they play, they ask each other questions, such as "*peuco*, where do you come from?," "from the rushes bog," "why do you come?," "to hunt your chicks," "hunt them, if you can!"

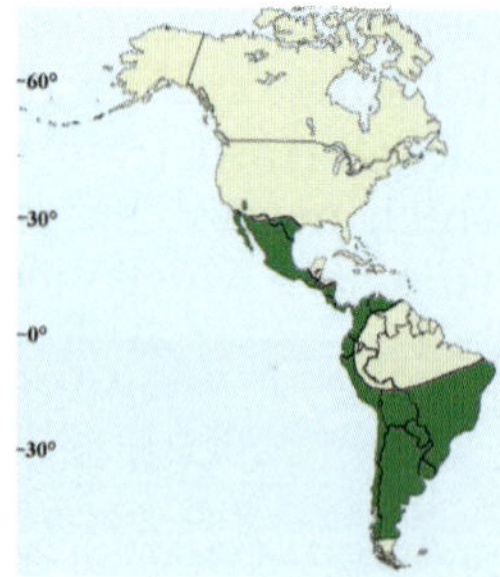

In spite of the stigmas it carries for being the "chicken-hunter", the Harris's Hawk, like other birds of prey, helps farmers and livestock owners by contributing to the control of rodent, bird and reptile populations in the countryside, shrublands and austral forests.

* *The Yahgan grandmothers Úrsula and Cristina Calderón recognize the Harris's Hawk in pictures, and ornithologist Steve McGehee observed one individual of* Parabuteo unicinctus *on Navarino Island. This represents a new record, which considerably extends the latitudinal range known for this species.*

Yahgan: Weziyau, Wăriáo
Mapudungun: **Mañke**, Manque
Spanish: **Cóndor**, Cóndor andino
English: **Andean Condor**
Scientific: *Vultur gryphus*

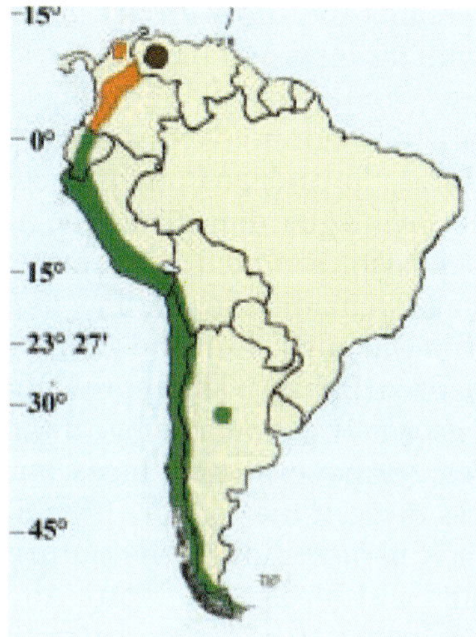

For the *Mapuche*, and for many other cultures, the Andean Condor or *mañke* is the king of birds. With a wingspan of more than 3 meters, it is the largest flying bird in the world. For the *Mapuche*, *mañke* also symbolizes the Andean Cordillera with its awesome size, its snow-white ring of neck feathers, and its black body, which suggests the Cordillera's rocks and minerals.

The king of birds flies at high altitudes and embodies the fundamental virtues of the Mapuche culture. The young man *Mañkian* derives his name from the expression *mañkean*, which means a man who wants to be like the Andean Condor (see the story of the Red-Backed Hawk). The surname *Mankelef* denotes a swift person or condor that maintains justice, who unites in himself the attributes of: *i) Kimche* or a wise person (*kim* = wisdom,*che* = person), *ii) Norche* or a person that loves justice (*nor* = rights), *iii) Kümeche* or a good-natured person (*küme* = good) and *iv) Newenche* or a powerful person, community ruler (*newen* = spiritual or physical strength and riches, like that of Caupolicán's valiant *longko* or Mapuche chief who gave his life for the freedom of his people).

Of all the birds in the world, the Andean Condor flies the highest, achieving altitudes that exceed 7,000 meters. The ancestral stories of the high elevation Andean cultures tell how *Viracocha* emerged from Lake Titicaca. *Viracocha* then created the sun with his light, the rain with his tears and the sky, the stars, humans, and other living beings that inhabit the region. With its majestic grandness, the Andean Condor or cuntur is *Viracocha's* guardian bird. The figure of *Viracocha*, sculpted in stone 2,200 years ago, remains silent today, looking towards the sunrise in the Sungate at Tiahuanaco (Bolivia). This representation of *Viracocha* is surrounded by 48 winged figures: 32 have human faces and 16 possess those of the condor. This figure reminds us how linked the natures of humans and birds are. Many Andean peoples believe to be descended from the cuntur. Like the *Mapuche*, their cosmology maintains that humans are connected by the Andean Condor to the high, alpine ecosystems with their rains, rivers and lakes, the heavens with the sun, moon, stars, soils, and the waters with all the living beings that inhabit them.

The modern nations of South America also admire the Andean Condor; Colombia, Ecuador, Bolivia, and Chile, have made it their "national bird." On the national shield of Chile the characteristic crest of the male condor is represented, tellingly, as a king's crown. Until the 18th century, this king of birds flew over the whole length of the Andean Cordillera from the Sierra de Santa Marta (Colombia) to Cape Horn. Throughout this range, the Andean Condor abounded from the peaks to the sea-level. As the poet Pablo Neruda writes, they are birds "born between the mountains and the marine spray." The tallest mountain in the capital of Chile is called Manquehue, which means "place of the condors"

(*mañke* = condor; *hue* = place). This indicates that the Andean Condor, *cuntur* or *mañke*, was abundant in the lands that today are occupied by large cities in the Andean region of South America.

Paradoxically, the king of birds, so admired by diverse cultures, has disappeared from many parts of its range. In the city of Mérida, Venezuela it was seen for the last time in 1912. In Quito, Ecuador it is necessary to travel hours on expeditions from the city to see the few condors that take refuge on the sides of Andean volcanoes. In Chile, *Vultur gryphus* is considered a species with serious conservation problems.

Andean Condors are particularly vulnerable species because they occupy large territories. They can travel up to 150 kilometers in a day to find carrion. They are also very long-lived organisms—living more than 50 years—and they begin to breed at the age of six years. They breed every two years, laying one or two eggs.

The risk that the Andean Condor may become extinct could alsoinvolve the disappearance of the fundamental virtues embodied by it: wisdom (*kim*), justice (*nor*), goodness (*küm*) and discipline (*newen*). Like these values that we yearn and reach for, the Andean Condor is frequently difficult to see because it often soars so high that it is not detectable by the naked eye. Fortunately, in the coastal forests of the extreme southern archipelagos of South America, the Andean Condor is still clearly observable while it descends to feed on litoral ecosystems. So, in these remote forests, *mañke* or *wariáo* astonishes us by its beauty and inspires a love for birds and for the highest values humans can cultivate during our brief, and shared life.

Mapuche story. The Condor or the *mañke* is for the Mapuche culture the King of the Birds. It is a symbol of the mountains, not only due to its great size and impressive wing-length, but also its black and white colors, which signify the snow-capped peaks and the black mountain rocks. For the Mapuche, the condor unites the virtues of being *Kimche*, a wise person, *Norche*, a person who loves justice, *Kümeche*, a good-natured person, and *Newenche*, a powerful, lordly person. This King of Birds is the national bird of Columbia, Ecuador, Bolivia and Chile. *Manke* was once abundant on *Manquehue* Mountain, which means place (*hue*) of the condors (*mañke*), in the center of Santiago, the capital of Chile. It is still found today throughout the territory of the austral forests, but paradoxically, this king of birds, this national symbol of four countries, finds itself now threatened with extinction.

Mapudungun version.
Affkappe Affkappe Affkappe Affkappe
Affkappe Affkappe Affkappe Affkappe
Tachi mañke fey ngenñidol puwerra üñümtuley pu mapuche tañi kimün meu. Pirentuku winkul püle rekeley. Fey tañi fütramongen ka feyñi lügkürü femngen. Feyta chi mañke kom fütra kim newen dungu wepümniyey kimche inanielu wera kimün, Norche nentunielu norum dungu, Kümeche llakonagduam, Newenche feyta ülmen ka fütra ñidol. Welulelureke, ngen ñidol puwera üñüm mülelu willi mapu püle cheuñi mülen mawidantu ka rumel pirentuku winkul tripan antüpüle komtrokiñ willi admapupüle, fey ñidol üñüm pingey Colombia, Ecuador, Bolivia ka Chülle. Mapu (Chile), kafey rume mülekefuy winkul Mankehue ñidol waria ta Chile, feula fey epeke ngewetulay larumelelu.

Yahgan: **Iloéa,** Īliūaia, Aiskalaix, Akixlaiix, Ilaóea
Mapudungun: **Kelüwún Kaniñ,** Queluy, Quelvoni, Kanin, Kelwi
Spanish: **Jote de cabeza colorada,** Gallinazo, Jote de cabeza roja
English: **Turkey Vulture**
Scientific: *Cathartes aura*

The Turkey Vulture is found on all of the continent and in a great variety of environments. From close up, it is distinguished by its red legs, head and neck. Its *Mapudungun* names *kelwi* or *kelüwún kaniñ*, allude to its red color (*kelü* = red, *Kaniñ* = vulture, *wün* = mouth and wünüñüm = beak; which is to say bird (*üñüm*) mouth (*wün*).

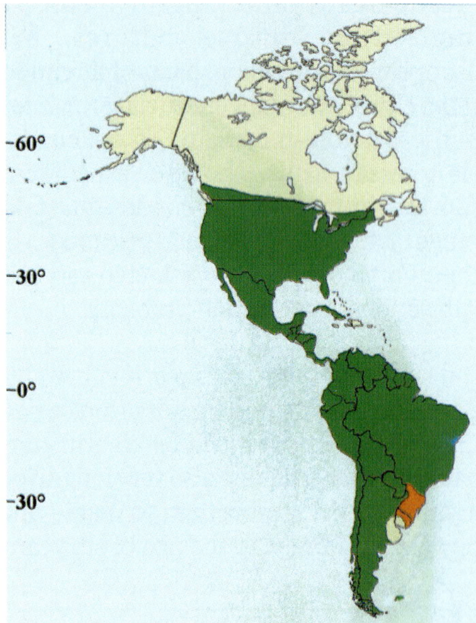

In flight, the lower half of the wings is translucent, even appearing white. Recently, it was discovered that these vultures have a very developed sense of smell that permits them to discern and locate decomposing meat from great distances. In its flights over temperate and tropical forests these carrion eaters have the abilit to detect food that is on the forest floor below the canopy.

When they nest, Turkey Vultures prefer to use low brushland or fallen trunks in the forest, and occupy the same nest for several years. In the Magellanic Archipelago Region, where the evergreen forests grow down to high tide line, Turkey Vultures are abundant in the littoral zone. In this forest-shore ecotone they nest in dense vegetation, and eat a great variety of carrion, especially in colonies of sea lions and penguins.

Vultures provide a great service to ecological systems by cleaning the bones of dead animals and contributing to the chain of decomposers that then turn dead organic matter into soil and new living beings.

For humans, they provide a special service by eliminating trash, and even taking away infectious bacteria. Due to the strong acidic conditions of vultures' stomachs, ingested bacteria are killed. In addition, when Turkey Vultures flight at high altitude, the intense ultraviolet radiation kills the bacteria adhered to their plumages. This ecological-cleaning service of vultures is poetically illustrated by the *Lafkenche* poet Lorenzo Aillapán, who in *Twenty Winged Poems* writes:

Janitor of the city, the vulture bird cleans.	*Tachi kaniñ üñüm liftu wariafe*
Favorably they comment, the strange ones who take notice.	*Pikeeyu ka pu che inarumengelu*
Full-time cleaner of the beach and the open countryside,	*Liftu küyüm rüpüfe ka feyti lelfünche*
The people of the place comment with admiration.	*Feypilekey pu che lofmeu*

Yahgan: out of bird´s range
Mapudungun: **Kaniñ kürüwún**, Queluy, Quelvoni
Spanish: **Jote de cabeza negra**
English: **Black Vulture**
Scientific: *Coragyps atratus*

The *Mapudungun* names of the vultures describe the color of the head of the two species that inhabit the southern forests of South America. As such, the Red-Headed (or Turkey) Vulture is *Kaniñ kelülonko* (*Kaniñ* = vulture; *kelü* = red; *lonko* = head) and the Black Vulture is *Kaniñ kürülonko* (*kürü* = black).

The Black Vulture inhabits the whole American continent, but does not reach the high latitudes that the Turkey Vulture does. In Chile and Argentina they reach a maximum southern range at 50° S, while the Turkey Vulture reaches Cape Horn (56° S).

Black Vultures are social birds and are frequently seen soaring in concentric circles, where groups of individuals cooperate in the search of a dead animal, using its sense of sight in open habitats. Its extended wings are completely black and its legs are also black and over-extend their short tail when in flight. In the forest edges of the austral forests it is common to observe groups of Black Vultures perched on trees, where they usually extend their wings to the sun to warm their body and to dry and put into line their feathers. To nest they use caves and rocky cliffs.

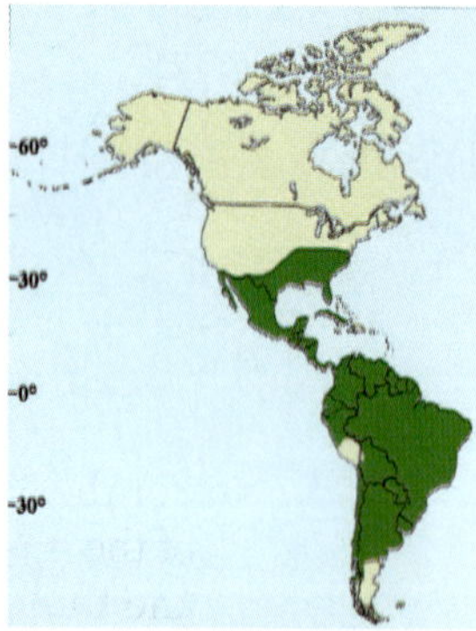

Although vultures prefer recently dead animals, they eat all types of carrion. Hence, vultures are seen as good cleaners. For this reason, when the *Huilliche* on Chiloé Island find an egg of vulture among their crops, they see in this a message of good luck, and they forecast good harvests.

The voices and stories must continue

For the grandparents and grandchildren
of the birds, *pij*, *üñüm* or *aves*
and the *che*, human beings, *yaganes* or *seres humanos*
that fly (or would like to fly)
as *omora*, *pinda*, hummingbird, *picaflor* or *Sephanoides sephaniodes*,
from flower to flower in the forests of austral South America,
feeding ourselves and feeding many other living beings,
captivated by the beauty, mystery and delicacy
of the odors, tastes, colors, textures and sounds
that emanate from this inexhaustible fount of biological and cultural diversity,
flying over the indifference and the dogmatism,
determined to live the risk and vitality of sharing the life with whom we co-inhabit
these rainy and fecund *forests, ashuna* or *mawida* of the planet,
we must continue listening and permitting the expression of their multiple voices.

Selected Bibliography

Ornithological references

Araya, B. & S. Chester. 1993. *The Birds of Chile*. Latour, Santiago, Chile.

Couve, E. & C. Vidal. 2000. *Aves del Canal de Beagle y Cabo de Hornos*. Fantástico Sur, Punta Arenas, Chile.

De la Peña, M. & M. Rumboll. 1998. *Birds of Southern South America and Antarctica*. HarperCollins Publishers, London.

De Schauensee, R. 1970. *A Guide to the Birds of South America*. Academy of Natural Sciences of Philadelphia by Livingston Pub. Co., Wynnewood, Pa.

Egli, G. & J. Aguirre. 2000. *Aves de Santiago*. UNORCH, Z&D Servicios Gráficos, Chile.

Elphick, C., J.B. Dunning Jr. & D. A. Sibley (eds.). 2001. *The Sibley Guide to Bird Life & Behavior*. National Audubon Society, Alfred A. Knopf, New York.

Fjeldså, J. & N. Krabbe. 1990. *Birds of the High Andes*. Zoological Museum and Apollo Books, Copenhagen and Stenstrup, Denmark.

Goodall, J.D., A.W. Johnson & R.A. Philippi. 1946. *Las Aves de Chile*. Volumen I. Establecimientos Gráficos Platt S.A., Buenos Aires, Argentina.

Goodall, J.D., A.W. Johnson & R.A. Philippi. 1951. *Las Aves de Chile*. Volumen II. Establecimientos Gráficos Platt S.A., Buenos Aires, Argentina.

Hellmayr, C.E. 1932. Birds of Chile. *Field Museum Natural History Publications 308 (Zoological Series)* 19: 1-472.

Hoffmann, A. & I. Lazo. 2000. *Aves de Chile. Un Libro También para Niños*. Universidad Andrés Bello, Ril Editores, Santiago.

Housse, R.E. 1945. *Las Aves de Chile en su Clasificación Moderna. Su Vida y Costumbres*. Ediciones Universidad de Chile, Santiago.

Jaksic F & P Feinsinger (1991) Bird assemblages in temperate forests of North and South America: a comparison of diversity, dynamics, guild structure and resource use. Revista Chilena de Historia Natural 64: 491-510.

Moreira, A. 1999. *Guía de Campo. Caleu y Cerro El Roble*. Asociación de Comuneros La Capilla de Caleu, Chile.

Narosky, T. & A. Bosso. 1995. *Manual del Observador de Aves*. Editorial Albatros Saci, Buenos Aires.

Narosky, T. & D. Yzurieta. 1987. *Guía para la Identificación de las Aves de Argentina y Uruguay*. Vazquez Mazzini Editores, Buenos Aires.

Ridgely, R.S. & G. Tudor. 1989. *The Birds of South America*. Volume 1, The Oscine Passerines. University of Texas Press, Austin.

Ridgely, R.S. & G. Tudor. 1994. *The Birds of South America*. Volume 2. University of Texas Press, Austin, Texas.

Rozzi, R., D. Martínez, M.F. Willson & C. Sabag. 1996. Avifauna de los Bosques Templados de Sudamérica. In *Ecología de los Bosques Nativos de Chile* (J.J. Armesto, C. Villagrán & M.T. Kalin, eds.), pp: 135-152. Editorial Universitaria, Santiago, Chile.

Venegas, C. 1994. *Aves de Magallanes*. Ediciones de la Universidad de Magallanes, Punta Arenas, Chile.

Vuilleumier, F. 1985. Forest birds of Patagonia: ecological geography, speciation, endemism, and faunal history. *Neotropical Ornithology, Ornithological Monographs* 36: 255-304.

Willson, M.F., T.L. de Santo, C. Sabag & J.J. Armesto. 1994. Avian communities of fragmented south-temperate rainforests in Chile. *Conservation Biology* 8:508-520.

References used in the preparation of bird species distribution range maps

Allen, T.B. (ed.). 1983. *Field Guide to the Birds of North America.* National Geographic Society, Washington.

Barros V.A. 1976. Nuevas aves observadas en las Islas Picton, Nueva, Lennox y Navarino oriental. *Anales del Instituto de la Patagonia* 7: 190-193.

Barros, A. 1971. Aves observadas en las Islas Picton, Nueva, Lennox y Navarino Oriental. *Anales del Instituto de la Patagonia* 2: 166-180.

Blake, E.R. 1977. *Manual of Neotropical Birds.* University of Chicago Press, Chicago.

Butler, T.Y. 1979. *The Birds of Ecuador and the Galapagos Archipelago: A Checklist of all the Birds Known in Ecuador and the Galapagos Archipelago and a Guide to Help Locate and See Them.* Ramphastos Agency, Portsmouth, N.H.

Davis L.I. 1972. *A Field Guide to the Birds of Mexico and Central America.* University of Texas Press, Austin.

De La Peña M. 1988. *Guía de Aves Argentinas. Columbiformes a Piciformes.* Facultad de Agronomía y Veterinaria, U.N.L.

De la Peña, R. & M. Rumboll. 1998. *Birds of Southern South America and Antarctica.* Collins Illustrated Checklist. HarperCollins*Publishers*, Rotolito Lombarda, Italy.

De Schauensee, R.M. & W.H. Phelps, Jr. 1978. *A Guide to the Birds of Venezuela.* Princeton University Press, Princeton, N.J.

De Schauensee, R.M. 1964. *The Birds of Colombia, and Adjacent Areas of South and Central America.* Livingston Pub. Co. Narberth, Pa.

De Schauensee, R.M. 1966. *The Species of Birds of South America and their Distribution.* Academy of Natural Sciences; dist. by Livingston, Narberth, Pa., Philadelphia.

del Hoyo, J., A. Elliott & J. Sargatal (eds). 1992. *Handbook of the Birds of the World. Volume 1: Ostrichs to Ducks.* Lynx Editions, Barcelona.

del Hoyo, J., A. Elliott & J. Sargatal (eds.). 1994. *Handbook of the Birds of the World. Volume 2: New World Vultures to Guineafowls.* Lynx Editions, Barcelona.

del Hoyo, J., A. Elliott & J. Sargatal (eds.). 1996. *Handbook of the Birds of the World. Volume 3: Hoatzin to Auks.* Lynx Edicions, Barcelona.

del Hoyo, J., A. Elliott & J. Sargatal (eds.). 1997. *Handbook of the Birds of the World. Volume 4: Sandgrouse to Cukoos.* Lynx Editions, Barcelona.

del Hoyo, J., A. Elliott & J. Sargatal (eds.). 1999. *Handbook of the Birds of the World. Volume 5: Barn Owls to Hummingbirds.* Lynx Editions, Barcelona.

del Hoyo, J., A. Elliott & J. Sargatal (eds.). 2001. Handbook of the Birds of the World. Volume 6: Mousebirds to Hornbills. Lynx Editions, Barcelona.

Dunning, J.S. 1982. *South American Land Birds.* Harrowood Book, Newtown Square, Pennsylvania.

Garrido, O.H. & F. García-Montaña. 1975. *Catálogo de las Aves de Cuba.* La Habana, Cuba.

Heintzelman, D.S. 1979. *Hawks and Owls of North America: A Complete Guide to North American Birds of Prey.* Universe Books, N.Y.

Johnsgard, P.A. 1990. *Hawks, Eagles and Falcons of North America.* The Smithsonian Institute Press, Washington D.C.

Land, H. 1970. *Birds of Guatemala*. Published for the International Committee for Bird Preservation, Pan-American Section by Livingston Pub. Co., Wynnewood, Pa.

Monroe, B.L. Jr. 1968. *A Distributional Survey of the Birds of Honduras*. Ornithological Monographs N°7. American Ornithologists' Union, N.Y.

Narosky, T. & M. Babarskas. 2000. *Aves de la Patagonia. Guía para su Reconocimiento*. Vásquez Mazzini Editores, Buenos Aires.

Olrog, C.C. 1963. *Lista y Distribución de las Aves Argentinas*. Tucumán, Argentina.

Parker, T.A. III, D.F. Stotz & J.W. Fitzpatrick. 1996. *Neotropical Birds: Ecology and Conservation*. University of Chicago Press, Chicago.

Paynter, R.A. & M.A. Traylor, Jr. 1977. *Ornithological Gazetteer of Ecuador*. Bird Dept., Museum of Comparative Zoology, Harvard University, Cambridge, Mass.

Paynter, R.A. & M.A. Traylor, Jr. 1981. *Ornithological Gazetteer of Colombia*. Bird Dept., Museum of Comparative Zoology, Harvard University, Cambridge, Mass.

Paynter, R.A. Jr. 1982. *Ornithological Gazetteer of Venezuela*. Bird Dept., Museum of Comparative Zoology, Harvard University, Cambridge, Mass.

Paynter, R.A. Jr. 1989. *Ornithological Gazetteer of Paraguay*. Bird Dept., Museum of Comparative Zoology, Harvard University, Cambridge, Mass.

Paynter, R.A., Jr., M.A. Traylor, Jr. & B. Winter. 1975. *Ornithological Gazetteer of Bolivia*. Bird Dept., Museum of Comparative Zoology, Harvard University, Cambridge, Mass.

Perrins, C.M. 1990. *The Illustrated Encyclopedia of Birds: The Definitive Reference to Birds of the World*. Prentice Hall Editions, N.Y.

Pinto, O.M.O. 1960. *Beija-flores do Brasil*. Livraria São José, Rio de Janeiro.

Rand, D.M. & R.A. Paynter, Jr. 1981. *Ornithological Gazetteer of Uruguay*. Bird Dept., Museum of Comparative Zoology, Harvard University, Cambridge, Mass.

Remsen, Jr, J.V. & M.A. Traylor, Jr. 1989. *An Annotated List of the Birds of Bolivia*. Buteo Books, Vermillion, South Dakota.

Ridgely, R.S. & J.A. Gwynne, Jr. 1989. A Guide to the Birds of Panama, with Costa Rica, Nicaragua, and Honduras. 2nd ed. Princeton University Press, Princeton, N.J.

Robbins, C.S., B. Bruun & H.S. Zim. 1983. *Birds of North America: A Guide to Field Identification*. Golden Press, N.Y.

Sibley, C.G. & B.L. Monroe, Jr. 1990. *Distribution and Taxonomy of Birds of the World*. Yale University Press, N.H.

Snyder, D.E. 1966. *The Birds of Guyana (formerly British Guinna); A Checklist of 720 Species, with Brief Descriptions, Voice and Distribution*. Peabody Museum, Salem.

Stephens, L. & M.A. Traylor, Jr. 1985. *Ornithological Gazetteer of the Guianas*. Bird Dept., Museum of Comparative Zoology, Harvard University, Cambridge, Mass.

Stiles, F.G. & A.F. Skutch. 1989. *A Guide to the Birds of Costa Rica*. Comstock, Ithaca.

Texera W (1972) Distribución y diversidad de mamíferos y aves en la Provincia de Magallanes. Análisis preliminar de la diversidad ecológica y variación taxonómica. Anales del Instituto de la Patagonia (Chile) 3: 171-200.

Texera W (1973) Distribución y diversidad de mamíferos y aves en la Provincia de Magallanes. II: Algunas notas ecológicas sobre los canales patagónicos. Anales del Instituto de la Patagonia (Chile) 4(1-3): 291-305.

Venegas, C. 1981. Aves de las Islas Wollaston y Bayly, Archipiélago del Cabo de Hornos. *Anales del Instituto de la Patagonia* 12: 213-219.

Venegas, C. 1991. Ensambles avifaunísticos estivales del Archipiélago Cabo de Hornos. *Anales del Instituto de la Patagonia* 20: 69-81.

Vigil, C. 1973. *Aves Argentinas y Sudamericanas*. Editorial Atlántida, Buenos Aires.
Woods, R.W. 1975. *The Birds of the Falkland Islands*. Anthony Nelson, Oswestry.
Zalles, J.I. & K.L. Bildstein (eds). 2000. *Raptor Watch: A Global Directory of Raptor Migration Sites*, Birdlife International, Cambridge.

References about ecology and conservation of South American temperate forests and their avifauna

Anderson, C. & R. Rozzi. 2000. Bird assemblages in the southernmost forests in the world: Methodological variations for determining species composition. *Anales del Instituto de la Patagonia* 28: 89-100.

Armesto, J.J, C. Villagrán & M.T. Kalin. 1996. *Ecología de los Bosques Nativos de Chile*. Editorial Universitaria, Santiago, Chile.

Armesto, J.J., R. Rozzi, C. Smith-Ramírez & M.T.K. Arroyo. 1998. Effective conservation targets in South American temperate forests. *Science* 282: 1271-1272.

Arroyo, M.T.K., M. Riveros, A. Peñaloza, L. Cavieres & A.M. Faggi. 1996. History and regional richness patterns of the cool temperate rainforest flora of Southern South America. En R.G. Lawford, P. Alaback & E.R. Fuentes (eds.), *High Latitude Rain Forest and Associated Ecosystems of the West Coast of the Americas: Climate, Hydrology, Ecology and Conservation*. Springer-Verlag, Berlin, pp. 135-151.

Axelrod, D.I., M.T.K. Arroyo & P. Raven. 1991. Historical development of temperate vegetation in the Americas. *Revista Chilena de Historia Natural* 64: 413-446.

Cofre H (1999) Patrones de rareza de las aves del bosque templado de Chile: implicancias para su conservación. Boletín Chileno de Ornitología 6: 8-16.

CONAF-CONAMA-BIRF, 1997. *Catastro y Evaluación de los Recursos Vegetacionales Nativos de Chile*. Universidad Austral de Chile, PUC, Universidad Católica de Temuco.

Donoso, C. 1995. *Bosques Templados de Chile y Argentina*. Editorial Universitaria, Santiago, Chile.

Erazo S (1987) Observaciones y alcances ecológicos a la comunidad de aves terrestres en la desembocadura del río Aconcagua, V Región, Chile. Revista Geográfica de Valparaíso (Chile) 18: 51-62.

Estades C & S Temple (1999) Deciduous-forest bird communities in a fragmented landscape dominated by exotic pine plantations. Ecological Applications 9: 573-585.

Fuentes, E. 1994. *¿Qué Futuro Tienen Nuestros Bosques?. Hacia la Gestión Sustentable del Paisaje del Centro y Sur de Chile*. Ediciones Universidad Católica de Chile, Santiago, Chile.

Gajardo, R. 1994. *La Vegetación Natural de Chile: Clasificación y Distribución Geográfica*, Editorial Universitaria, Santiago, Chile.

Glade A (ed) (1988) Libro rojo de los vertebrados chilenos. Corporación Nacional Forestal, Santiago (Chile).

Hinojosa LF & Villagrán, C. 1997. Historia de los bosques del sur de Sudamérica, I: antecedentes paleobotánicos, geológicos y climáticos del Cono Sur de América. *Revista Chilena de Historia Natural* 70: 225-239.

Jaksic F (1997) Ecología de los vertebrados de Chile. Ediciones Universidad Católica de Chile. Santiago, Chile. 262 pp.

López, M.V. 1990. Variación estacional en el uso de los recursos alimenticios para algunos componentes de una taxocenosis de aves Passeriformes en Quebrada de la Plata, Chile Central. Tesis de Magister. Facultad de Ciencias, Universidad de Chile. 116 pp.

Muñoz, M., H. Nuñez & J. Yánez (comps.). 1996. *Libro Rojo de los Sitios Prioritarios para la Conservación de la Diversidad Biológica en Chile*, Ministerio de Agricultura-CONAF, Santiago, Chile.

Primack, R., R. Rozzi, P. Feinsinger, R. Dirzo & F. Massardo. 2001. *Fundamentos de Conservación Biológica: Perspectivas Latinoamericanas*. Fondo de Cultura Económica, México.

Rau J, A Gantz & G Torres (2000) Estudio de la forma de fragmentos boscosos sobre la riqueza de especies de aves al interior y exterior de áreas silvestres protegidas. Gestión Ambiental 6: 33-40.

Riveros G.M. & M.C. Lopez-Calleja 1990. Distribución de las aves en el período no reproductivo y su relación con las formaciones vegetacionales presentes en el Parque Nacional la Campana, Chile Central. *Boletín Sociedad de Biología de Concepción* 61:161-166.

Rodríguez, R., O. Matthei & M. Quezada. 1983. *Flora Arbórea de Chile. Biblioteca de Recursos Renovables y no Renovables de Chile*. Editorial Universidad de Concepción, Concepción, Chile.

Rozzi, R., J.J. Armesto, A. Correa, J.C. Torres-Mura & M. Sallaberry. 1996. Avifauna de bosques primarios templados en islas deshabitadas del archipiélago de Chiloé. Revista Chilena de Historia Natural 69: 125-139.

Schlatter R (1976) Contribución a la ornitología de la Provincia de Aysén. Boletín Ornitológico (Chile) 8 (1): 3-18.

Silander, J.A., Jr. 2000. Temperate forests: plant species biodiversity and conservation, In *Encyclopedia of Biodiversity*, S.A. Levin, ed., Academic Press, New York, pp. 5:607-626.

Simonetti, J. A., M. T. K. Arroyo, A. E. Spotorno & E. Lozada (comps.). 1995. *Diversidad Biológica de Chile*, CONICYT, Santiago de Chile.

Tala C (1991) Avifauna observada en la Reserva Nacional Río de los Cipreses (VI Región). Boletín Informativo Unión de Ornitólogos de Chile (UNORCH) (Chile) 12: 13-15.

Tecklien D, A Vila & S Palminteri. 2003. A Biodiversity Vision for the Valdivian Temperate Rain Forest Ecoregion of Chile and Argentina. Informe Conservation Science Program, WWF, USA.

Torres-Mura JC (1994) Estado de conservación de la fauna terrestre de Chile. En: Espinoza G, P Pisani, L Contreras & P Camus (Eds.) Perfil Ambiental de Chile: 367-375. Comisión Nacional del Medio Ambiente, Santiago. 569 pp.

Veblen, T., R.S. Hill & J. Read (eds.). 1996. *The Ecology and Biogeography of Nothofagus Forests*, Yale University Press.

Veblen, T.T., F.M. Schlegel & J.V. Oltremari. 1983. Temperate broad-leaved evergreen forests of South America. En Ovington, J.D. (ed.), *Temperate Broad-leaved Evergreen Forests*. Ecosystems of the World Volume 10. Elsevier, Amsterdam, pp. 5-31.

Villagrán, C. & L. F. Hinojosa. 1997. Historia de los bosques del sur de Sudamérica, II: análisis fitogeográfico. *Revista Chilena de Historia Natural* 70:24 1-267.

Vuilleumier F (1985) Forest birds of Patagonia: ecological geography, speciation, endemism and faunal history. En: Buckley PA, MS Foster, ES Morton, RS Riedely & FG Buckley (eds) Neotropical Ornithology. Ornithological Monographs 36: 255-304.

Wilcox, K. 1995. *Chile's Native Forests: A Conservation Legacy*. Ancient Forest International, Redway, CA.

References concerning ethnography and biocultural conservation in South American temperate forests

Augusta, F. F. J. d. 1903. *Gramática Mapuche Bilingüe*. Ediciones Séneca, Santiago, Chile.

Aillapan, L. & R. Rozzi. 2001. *Veinte Poemas Alados de los Bosques Nativos del Sur de Chile*. Editorial Plaza y Valdés, México.

Arnold, J. 1996. The inverse system in Mapudungun and other languages. *Revista de Lingüística Teórica y Aplicada* 34.

Bart, F. 1948. Cultural development in Southern South America: Yahgan and Alakaluf vs. Ona and Tehuelche. *Acta Americana* 6: 192-9.

Bate, L. 1983. Comunidades primitivas de cazadores recolectores en Sudamérica. *Historia General de América*, Volumen 2. Caracas, Venezuela.

Bird, J. 1946. The Archaeology of Patagonia. En *Handbook of South American Indians*, Smithsonian Institution, Washington, USA.

Bridges, E.L. 1949. *Uttermost Part of the Earth*. E.P. Dutton and Company Inc., N.Y. (Special ed. 1950).

Bridges, T. 1933. *Yamana-English Dictionary*. F. Hestermann & M. Gusinde (Eds.). Second Edition (1987) Zagier y Urruty Publicaciones, Buenos Aires.

Cárdenas, R. 1994. *Diccionario de la lengua y de la Cultura de Chiloé*. Olimpho, Santiago.

Catrileo, M. 1988. *Mapudunguyu: Curso de Lengua Mapuche*. Universidad Austral de Chile.

Catrileo, M. 1998. *Diccionario Linguístico-Etnográfico de la Lengua Mapuche. Mapudungun-Español-English*. Editorial Andrés Bello, Santiago de Chile. 3°ed.

Citarella, L. (ed.). 1995. *Medicinas y Culturas en la Araucanía*. Editorial Sudamericana, Santiago, Chile.

Coña, P. & Ernesto Wilhelm de Moesbach (1930). *Vida y Costumbres de los Indígenas Araucanos en la Segunda Mitad del Siglo XIX* (Pascual Coña & Ernesto Wilhelm de Moesbach). Imprenta Cervantes, Santiago, Chile.

Cooper, J. 1946a. The Ona. En *Handbook of South American Indians*. Volume 1. Smithsonian Institution, Washington D.C., pp. 107-125.

Cooper, J.M. 1946. The Yahgan. En *Handbook of South American Indians. The Marginal Tribes*. Vol 1. J. H. Steward ed. Smithsonian Institution, Bull. 143. U.S. Government Printing Office, Washington. pp. 81-106.

Croese, R. (1980). *Estudio Dialectológico del Mapuche*. Universidad Austral de Chile, Valdivia.

Chapman, A. 1986. *Los Selk'nam. La vida de los Ona*. Emecé Editores, Buenos Aires.

Charno, J., L. Blumenfeld, D. Lewiston & B. Wentz. 1993. *Voices of Forgotten Worlds. Traditional Music of Indigenous People*. Ellipsis Arts, N.Y.

Dannemann, M., & A. Valencia. 1989. *Grupos Aborígenes Chilenos. Su Situación Actual y Distribución Territorial*. Universidad de Santiago de Chile, Instituto de Investigaciones del Patrimonio Territorial de Chile, Colección Terra Nostra Nº 15.

Darwin C. 1838. *The Voyage of the Beagle*. Reprint, London: Everyman's Library, 1975.

Darwin C. 1871. *The Descent of Man*. Princeton University Press, ed. (1981). Princeton, New Jersey.

Emperaire, J., 1963. *Los Nómades del Mar*. Ediciones de la Universidad de Chile, Santiago de Chile.

Erize, E. 1960. *Diccionario Comentado Mapuche-Español*. Cuadernos del Sur, Universidad Nacional del Sur, Buenos Aires, Argentina.

Gorgoglione, E.C. 1997. Guía de Campo para la Identificación de las Aves del Neuquen. 1°ed. Editorial Hemisferio Sur, Montevideo, Uruguay.

Grimes, B. (ed.). 2000. ETHNOLOGUE: Languages of the World. 14th Edition, Summer Institute of Linguistics, Dallas, Texas.

Gusinde, M. 1946. *Urmenschen im Feuerland*. Paul Zsolnay Verlag, Berlin.

Gusinde, M. 1961. *The Yamana: The Life and Thought of the Water Nomads of Cape Horn*. Volumes I-V, translated by F. Schutze. New Haven Press, USA.

Gusinde, M. 1968. *Expedición a Tierra del Fuego*. Editorial Universitaria, 2° edición, Imprenta Salesianos, Santiago.

Hernández, A., N. Ramos & C. Cárcamo. 1997. *Diccionario Ilustrado Mapudungun Español Inglés*. Pehuen, Santiago, Chile.

Hidalgo, J., V. Schiappacasse, H. Niemeyer, C. Aldunate & P. Mege (eds.). 1996. *Etnografía. Sociedades Indígenas Contemporáneas y su Ideología*. Editorial Andrés Bello, Santiago, Chile.

Hyades, P. 1887. *Mission Scientifique du Cap Horn 1882-1883*, Tomo VII "Anthropologie et ethnographie". Paris, Gauthier-Villars.

King, Captain Parker. 1839. *Narrative of the Surveying Voyages of His Majesty's Ships Adventure and Beagle, Between the Years 1826 and 1836*. Henry Colburn. V 1., London.

Koppers, W. 1997. Entre los Fueguinos. Ediciones de la Universidad de Magallanes, Punta Arenas, Chile.

Latcham, R. 1928. *La Prehistoria Chilena*. Editorial Universo, Santiago de Chile.

Legoupil, D. 1993. El Archipiélago de Cabo de Hornos y la costa sur de la Isla Navarino: poblamiento y modelos económicos. *Anales del Instituto de la Patagonia* 22: 101-121.

Lenz, R. 1905-1910. *Diccionario Etimolójico de las Voces Chilenas Derivadas de las Voces Indíjenas Americanas*. Imprenta Cervantes, Santiago, Chile.

McEwan, L.A. Borrero & A. Prieto (eds.). 1997. *Patagonia: Natural History, Prehistory and Ethnography at the Uttermost End of the Earth*, Princeton University Press.

Mena, F. 1988. Cazadores recolectores en el área patagónica y tierras bajas aledañas (Holoceno medio y tardío). *Revista de Arqueología Americana* 4: 134-167.

Mires, A. 2000. *Así en las Flores como en el Fuego. La Deidad Colibrí en Amerindia y el Dios Alado en la Mitología Universal*. Ediciones Abya-Yala, Quito, Ecuador.

Nichols, J. 1990. Linguistic diversity and the first settlement of the New World. *Language* 66:475-521.

Ortiz-Troncoso, O. 1996. Los últimos canoeros. En Hidalgo J., V. Schiappacase, H. Niemeyer, C. Aldunate and P. Mege (eds.). *Etnografía. Sociedades Indígenas Contemporáneas y su Ideología. Culturas de Chile*. Editorial Andrés Bello, Santiago de Chile. pp.135-147.

Piana, E.L., A. Vila, L. Orquera, & J. Estevez. 1992. Chronicles of Onashaga: Archaeology in the Beagle Channel. *Antiquity* 66:771-783

Plath, O. 1996. Lenguaje de los Pájaros Chilenos. (2°). 1ª edición en Editorial Grijalbo. Santiago de Chile.

Ramirez-Sanchez, C. 1989. *Voces Mapuches*. Valdivia, Chile.

Rivas P., C. Ocampo & E. Aspillaga. 1999. Poblamiento temprano de los Canales Patagónicos: el núcleo septentrional. *Anales del Instituto de la Patagonia* 27: 221-230.

Roquera, L. & E. Piana. 1999. *La Vida Social y Material de los Yamana*. EUDEBA, Buenos Aires, Argentina.

Rozzi, R., J. Silander, J.J. Armesto, P. Feinsinger & F. Massardo. 2000. Three levels of integrating ecology with the conservation of South American temperate forests: The initiative of the Institute of Ecological Research Chiloé, Chile. *Biodiversity and Conservation* 9: 1199-1217.

Rozzi, R. 1999. The reciprocal links between evolutionary-ecological sciences and environmental ethics. BioScience 49 (11): 911-921.

Rozzi, R. 1998. *La filosofía ambiental de Callicott: entre un multiculturalismo y una ética ecocéntrica universal.* In "Los Caminos de la Etica Ambiental: Una Antología con Textos Contemporáneos» (Kwiatkowska, T. & J Issa, eds), pp.: 79-85. Editorial Plaza y Valdés, Universidad Autónoma Metropolitana Iztapalapa, Ciudad de México, México.

Rozzi, R. 2001. *Éticas ambientales latinoamericanas: raíces y ramas.* En "Fundamentos de Conservación Biológica: Perspectivas Latinoamericanas", Primack, R., R. Rozzi, P. Feinsinger, R. Dirzo & F. Massardo, pp. 311-362. Fondo de Cultura Económica, México.

Rozzi, R. 2002. Biological and Cultural Conservation in the Archipelago Forest Ecosystems of Southern Chile. Ph.D. Dissertation, Department of Ecology and Evolutionary Biology, University of Connecticut, USA.

Rozzi R. 2003. *Biodiversity and Social Wellbeing in South America.* In "Encyclopedia of Life Support Systems" (EOLSS). UNESCO-EOLSS. http://www.eolss.net

Rozzi R., F. Massardo, J. Silander Jr., C. Anderson & A. Marin. 2003. Conservación biocultural y ética ambiental en el extremo austral de América: oportunidades y dificultades para el bienestar ecosocial. In "Biodiversidad y Globalización" (Figueroa E & J Simonetti, eds.), pp. 51-85. Editorial Universitaria, Santiago, Chile.

Salas, A. 1984. *Textos Orales en Mapuche o Araucano del Centro-Sur de Chile.* Editorial de la Universidad de Concepción, Concepción, Chile.

Salas, A. 1992. Lingüística Mapuche. Guía Bibliográfica. *Revista Andina* Año 10 (N°2).

Smeets, I. 1989. *A Mapuche Grammar.* Leiden, Holland.

Stambuk, P. 1986. *Rosa Yagán. El Ultimo Eslabón.* Editorial Andrés Bello, Santiago, Chile.

Villagran, C., R. Villa, L. Hinojosa, G. Sánchez, M. Romo, A. Maldonado, L. Cavieres, C. Latorre, J. Cuevas, S. Castro, C. Papic & A. Valenzuela. 1999. Etnozoología Mapuche: un estudio preliminar. *Revista Chilena de Historia Natural* 72: 595-628.

Wilhelm de Moesbach, P. E. 1991. *Idioma Mapuche.* Imprenta y Editorial San Francisco, Villarica, Chile.

Authors Addresses

Lorenzo Aillapan
Poet, Mapuche Bird Man
Academia Mapuche
Püllümapukimunweftuy.
Puerto Saavedra, IX Región, Chile.

Christopher Anderson
Ecologist, Ph.D.(c)
Omora NGO, Puerto Williams, Chile
Institute of Ecology, University of Georgia (EEUU)
Email: cba@uga.edu

Úrsula Calderón
Crafts-woman
Comunidad Indígena Yagán de Bahía Mejillones.
She lived on Navarino Island, Chile, until January 2003. Today she rests on the island, in the Cemetery of Mejillones.

Cristina Calderón
Crafts-woman
Comunidad Indígena Yagán de Bahía Mejillones.
Villa Ukika, Puerto Williams, Isla Navarino, Chile.

George A. Clark
Ornithologist, Ph.D.
Emeritus, Department of Ecology & Evolutionary Biology University of Connecticut, EEUU.

Guillermo Egli
Teacher and Ornithologist
Unión Ornitológica de Chile (UNORCH).
Casilla 13183, Santiago 21, Chile.

Ezio M. Firmani
Chef and Producer
Omora NGO, Puerto Williams, Isla Navarino, Chile.

Uta Giesen
Geographer M.S. NGO Omora, Puerto Williams, Chile.
Email: uta.meike@gmx.de

Luis Gómez
Teacher
Comunidad Indígena Yagán de Bahía Mejillones, Villa Ukika, Isla Navarino Chile.
Quellón, Chiloé, X Región, Chile.

Nolberto González
Music teacher
Estudios Button Records, Puerto Williams, Chile.
Email: buttonrecords@hotmail.com

Kurt Heidinger
Writer Ph.D. NGO Omora & University of Connecticut, USA.
Email: lichtme@hotmail.com

Francisca Massardo
Etnobiologist, Ph.D.
Omora NGO & Universidad de Magallanes
Puerto Williams, Chile.
Email: francisca.massardo@umag.cl

Steven McGehee
Ornithologist
ONG Omora, Puerto Williams, Chile.
Email: whitethroatedcaracara@yahoo.com

Lorena Peñaranda
Lawyer
Omora NGO
Viña del Mar, Chile.
Email: juridicodirsomar@directemar.cl

Eduardo Ramilo
Ornithologist
Delegación Regional Patagonia
Administración de Parques Nacionales
Casilla 380 / San Carlos de Bariloche, Argentina.

Ricardo Rozzi
Philosopher and Ecologist, Ph.D.
Omora NGO & Universidad de Magallanes
Email: ricardo.rozzi@umag.cl

John Schwenk
Recording Engineer
WHUS 91.7 FM – University of Connecticut, EEUU.
Email: chiefengineer@whus.org, johnschwenk@juno.com

Oliver Vogel
Photographer
ONG Omora, Puerto Williams, Chile.
Email: foto-olv@web.de

Cristina Zárraga
Poet and writer
Comunidad Indígena Yagán de Bahía Mejillones
ONG Omora, Puerto Williams, Chile.
Email: hannuja@yahoo.com

Photographic Credits

Oliver Vogel
pages: 12(recording with Aillapan at Omora Park), 16(Button Records Studio), 19(Deciduous Beech forest), 32(Grandmother Cristina Calderón), 38 *(Campephilus magellanicus* - trunk), *40 (Campephilus magellanicus* -soil), *47(Pygarrhichas albogularis), 49 (Aphrastura spinicauda), 69(Cinclodes fuscus), 87 (Phrygilus patagonicus), 87(Phrygilus patagonicus), 93(Anairetes parulus), 95(Xolmis pyrope)*

Eduardo Ramilo
pages: 57(Bubo magellanicus), 66(Pardirallus sanguinolentus), 68(Cinclodes fuscus), 73(Pygochelidon cyanoleuca), 74(Theristicus melanopis), 77 (Vanellus chilensis), 80(Zenaida auriculata), 84(Elaenia albiceps), 86 (Carduelis barbatus), 94(Zonotrichia capensis), 96 (Curaeus curaeus), 107 (Milvago chimango), 112(Geranoaetus melanoleucus), 114(Buteo polyosoma), 117(Vultur gryphus)

Ricardo Rozzi
pages: 12(Workshop in Cucao), 14(Recording group in Mejillones), 15 (WHUS Studio), 17*(Araucaria araucana),* 19(Valdivian understory), 20 (Subantarctic forests), 31(Cristina and Úrsula Calderón), 33 (Aillapan at Omora Park Entrance), *108(Caracara plancus)*

Steve Morello
pages: 42(Scytalopus magellanicus), 43(Pteroptochos tarnii), 44(Scelorchilus rubecula), 45(Scelorchilus rubecula), 70(Cinclodes patagonicus), 85(Elaenia albiceps), 91(Troglodytes aedon), 98 (Sephanoides sephaniodes nesting), 105(Phytotoma rara), 122(Coragyps atratus)

Enrique Couve
pages: 53(Sylviorthorhynchus desmursii), 79(columba araucana), 82(Colaptes pitius), 83(Picoides lignarius), 90(Troglodytes aedon), 92(Colorhamphus parvirostris), 104(Sturnella loyca)

Eduardo Pavez
pages: 59(Tyto alba), 61(Accipiter bicolor), 115(Buteo polyosoma), 116(Parabuteo unicinctus), 118(Vultur gryphus)

Michel Sallaberry
pages: 52(Enicognathus ferrugineus), 67(Pardirallus sanguinolentus), 81(Zenaida auriculata), 89 (Turdus falcklandii), 110 (Caracara plancus), 121(Cathartes aura)

Steven McGehee
pages: 60 (Glaucidium nanum), 65(Gallinago gallinago), 72(Tachycineta leucopyga)

Günter Ziesler
pages: 97(Sephanoides sephaniodes), 98(Sephanoides sephaniodes flying), 99(Sephanoides sephaniodes), 101(Patagona gigas)
Joan Morrison
pages: 106(Milvago chimango), 109(Caracara plancus)
Ricardo Stanoss
pages: 20(Evergreen Magellanic forest), *62(Ceryle torquata), 63(Ceryle torquata)*
Christopher Anderson
pages: 48(Aphrastura spinicauda), 51(Enicognathus ferrugineus)
Augustin Berghöfer
15 (Recording on Navarino Island)
Alejandro Correa
39 (Campephilus magellanicus female)
Gonzalo E. González
102 (Diuca diuca)
Jaime Jiménez
78 (Columba araucana)
Alejandro Kusch
75 (Theristicus melanopis nesting)
Francisca Massardo
19 (Valdivian rainforest)
Patricio Riquelme
88(Turdus falcklandii)
Ricardo Rozzi Ricotti
113 (Geranoaetus melanoleucus)
John Schwenk
13 (Recording with Úrsula Calderón)
Sebastián Spongia
111 (Falco sparverius)
Juan Carlos Torres-Mura
55 (Strix rufipes)
María Angélica Vukasovic
54 (Strix rufipes nesting)
Norm Wickett
54 (Strix rufipes in a tree)

Drawings
Jay Barry: caricatures of Humberto Maturana, Charles Darwin, and hummingbird (*Sephanoides sephaniodes*)
Daniel Martínez: Ruffous-Collared Sparrow (*Zonotrichia capensis*), hummingbird (*Sephanoides sephaniodes*), and barn owl (*Tyto alba*)
Figures
Francisca Massardo: worldmap (p.17), map of forest types (p.18), ethnographic map (p.22), multi-ethnic scheme (p.23), hummingbird-bumblebee scheme (p.24), caracara transformation (p.26), and woodpecker transformation (p.26).

Stories´voices

Yahgan
Cristina Calderón: Green-Backed Hummingbird, Buff-Necked Ibis, Magellanic Tapaculo, White-Throated Treerunner.
Úrsula Calderón: Fire-Eyed Diucon, Common Snipe, Black-Chinned Siskin, Barn Owl, Thorn-Tailed Rayadito, Magellanic Woodpecker, Austral Trush.

Mapudungun
Lorenzo Aillapan

English
Kurt Heidinger: Chucao Tapaculo, Ringed Kingfisher, Swalow, White-Crested Elaenia, Striped Woodpecker, Eared Dove, Patagonian Tyrant, Long-Tailed Meadowlark, Southern Lapwing, Chimango Caracara, Southern-Crested Caracara, Red-Backed Hawk, Andean Condor, Green-Backed Hummingbird, Buff-Necked Ibis, Magellanic Tapaculo, Magellanic Woodpecker.
Uta Giesen: Fire-Eyed Diucon, Common Snipe, Black-Chinned Siskin, Barn Owl, Thorn-Tailed Rayadito, White-Throated Treerunner, Austral Trush.
Steven McGehee: Rufous-Legged Owl, Plumbeous Rail, Chilean Flicker, Austral Trush.

Index of Yahgan Bird Names

Index of Mapudungun Bird Names

Index of Spanish Bird Names

Index of English Bird Names

Index of Scientific Bird Names

Multi-ethnic Bird Guide

of the Austral Temperate Forests of South America

CD-I

		Track	Name	Time
Introduction		**1**	Trutruka song (Lorenzo Aillapan)	1:06:51
Bird Names	Forest Interior	**2**	Magellanic Woodpecker	0:40:57
		3	Magellanic Tapaculo	0:22:36
		4	Black-Throated Huet-Huet	0:32:65
		5	Chucao Tapaculo	0:31:14
		6	White-Throated Treerunner	0:24:40
		7	Thorn-Tailed Rayadito	0:25:40
		8	Austral Parakeet	0:28:57
		9	Desmur´s Wiretail	0:50:67
	Owls	**10**	Rufous-Legged Owl	0:43:14
		11	Austral Great Horned Owl	0:31:21
		12	Barn Owl	0:27:52
		13	Austral Pygmy Owl	0:28:16
		14	Bicolored Hawk	0:17:52
	Aquatics	**15**	Ringed Kingfisher	0:31:60
		16	Common Snipe	0:39:17
		17	Plumbeous Rail	0:29:27
		18	Bar-Winged Cinclodes	0:25:18
		19	Dark-Bellied Cinclodes	0:24:02
		20	Chilean Swalow	0:31:26
		21	Blue-and-White Swalow	0:25:24
		22	Buff-Necked Ibis	0:33:21
		23	Southern Lapwing	0:23:18
	Forest Edge	**24**	Chilean Pigeon	0:24:74
		25	Eared Dove	0:31:08
		26	Chilean Flicker	0:20:35
		27	Striped Woodpecker	0:29:28
		28	White-Crested Elaenia	0:25:74
		29	Black-Chinned Siskin	0:24:68
		30	Patagonian Sierrafinch	0:29:02
		31	Austral Thrush	0:26:34
		32	House Wren	0:30:14
		33	Patagonian Tyrant	0:21:39
		34	Tufted Tit-Tyrant	0:30:09
		35	Rufous-Collared Sparrow	0:39:12
		36	Fire-Eyed Diucon	0:35:61
		37	Austral Blackbird	0:35:07
		38	Green-Backed Firecrown	0:19:11
		39	Giant Hummingbird	0:20:06

Multi-ethnic Bird Guide

of the Austral Temperate Forests of South America

CD-I

		Track	Name	Time
Bird Names		40	Common Diuca Finch	0:31:18
		41	Long-Tailed Meadowlark	0:27:28
		42	Rufous-Tailed Plantcutter	0:30:19
	Raptors	43	Chimango Caracara	0:30:68
		44	Southern-Crested Caracara	0:27:17
		45	American Kestrel	0:22:10
		46	Black-Chested Buzzard-Eagle	0:22:22
		47	Red-Backed Hawk	0:24:48
		48	Bay-Winged Hawk	0:20:68
		49	Andean Condor	0:23:33
		50	Turkey Vulture	0:26:74
		51	Black Vulture	0:29:17
Mapuche Stories		52	Chucao Tapaculo	2:50:15
		53	Rufous-Legged Owl	1:53:06
		54	Ringed Kingfisher	1:21:03
		55	Plumbeous Rail	1:54:16
		56	Swalow	1:49:55
		57	White-Crested Elaenia	2:44:42
		58	Chilean Flicker	1:29:31
		59	Striped Woodpecker	1:33:01
		60	Eard Dove	3:03:12
		61	Patagonian Tyrant	2:28:26
		62	Long-Tailed Meadowlark	2:09:36
		63	Southern Lapwing	2:46:69
		64	Chimango Caracara	1:09:38
		65	Southern-Crested Caracara	1:15:70
		66	Red-Backed Hawk	1:47:55
		67	Andean Condor	2:48:22
		Total		58:01

Multi-ethnic Bird Guide
of the Austral Temperate Forests of South America

CD-II

	Track	Name	Time
Yahgan Stories	1	*Omora* Green-Backed Hummingbird	6:58:69
	2	Fire-Eyed Diucon	1:58:07
	3	Buff-Necked Ibis	4:38:66
	4	Common Snipe	2:27:01
	5	Black-Chinned Siskin	5:05:50
	6	Barn Owl	2:17:13
	7	Thorn-Tailed Rayadito	1:25:58
	8	Magellanic Tapaculo	4:47:49
	9	White-Throated Treerunner	2:01:11
	10	Magellanic Woodpecker	4:25:41
Yahgan / *Mapuche Story*	11	Austral Trush	4:03:51
Piccolo Song	12	Gli Uccellini (comp. Andrés Alcalde, int. Guillermo Lavado)	1:52:56
		Total	42:05:00